Ripley's

Believe It or Not!®

Book of
Chance

CONSTANTIN NIKOPOULOS (1786-1841)
a Greek tutor who lived in Paris, France,
ALWAYS SAID HE COULD NEVER UNDERSTAND WHY A MAN
WOULD DIE FOR LOVE OF A WOMAN, BUT COULD APPRECIATE A
MAN DYING FOR LOVE OF A GOOD BOOK
*DUSTING ONE OF HIS BELOVED VOLUMES, HE STRUCK
IT AGAINST HIS SLEEVE
- AND BY CHANCE INFLICTED HIS DEATH !*
AND THE MORAL IS: IF YOU LIKE THIS THIS ONE, YOU'LL LOVE

THE BOOK OF CHANCE

Ripley's

Believe It or Not!®

BOOK
OF
CHANCE

Ripley Books
Toronto/New York/London

Library of Congress Cataloguing in Publication Data

Main entry under title:

Ripley's Believe It or Not! : Book of Chance

   1. Curiosities and wonders.
AG243.R45   031'.02   82-7419
ISBN 0-698-11197-4   AACR2

Published in the United States by
Coward, McCann & Geoghegan, Inc.
200 Madison Avenue, New York,
N.Y. 10016, U.S.A.

Printed in the United States of America

Editor	Patrick G. Crean
Associate Editor	John McCuaig
Writers	Robert Crew
	David Miller
Assistant Editor	Gail Muir
Ripley Archivist	Edward Meyer
Researchers	Lydia Foy
	Judy MacGregor-Smith
	Victoria McCuaig
	Michael Johnson
Photographs	Mary Corcoran
	Julia Mustard
Design	Maher & Murtagh
Design assistant	Holly Fisher
Index	Edna Barker

Table of Contents ★ ★ ★ ★ ★ ★ ★

Introduction: The Chance of a Lifetime

Part One:

It's Show Time! Show Biz and the Big Break 2

* chance discoveries of movie stars
* career successes and failures
* quirks of fate and famous lives
* missed opportunities and sheer coincidence
* the world of music and the lucky break
* great writers and the vagaries of publishing
* famous artists and the whims of fortune
* frauds and fakes

Part Two:

The Wheel of Fortune Bets, Contests and the Sporting Life 42

* playing the odds
* lotteries, cards, horse racing, dice
* luck of the draw: fortunes won and lost
* casinos
* contests and lotteries: the winners
* superstitions in sports and gaming
* betting advantages in sports
* eccentric bets

Part Three:

Rich Man, Poor Man Free Enterprise and the Lucky Break 84

* rags to riches: famous fortunes
* hidden treasures
* boom and bust: fickle markets
* the world's richest people
* insurance risks
* missing heirs
* bright long-shot ideas and why they worked
* best chances of getting rich

★ ★ ★ ★ ★ ★ ★ ★ ★ ★ ★ ★ ★ ★ ★ ★

Part Four:

Blind Forces
Hazards, Marvels of
Invention and the Natural World
120

* chance and evolution
* survivors of natural disasters
* amazing technological inventions
* the chances of man's survival on earth
* fate and the origins of life
* chances of life in outer space
* paranormal happenings in nature
* possibilities for the future

Part Five:

Tangled Web
Crime and the Scales of Justice
156

* odds of being a victim
* the elimination of detective guesswork
* retribution by the dead
* convicted but not guilty
* crime statistics
* off-beat crimes
* innocent bystanders, mistaken identity
* clues found by accident, criminals caught by chance

Part Six:

A Day in Your Life
Health, Happiness
and Your Daily Risks
176

* marriage and coincidence
* common accidents
* sleep, suicide and disease
* genetics, genius and heredity
* twins and the chances of simply being born
* medical marvels and inventions
* anatomical oddities and miracles
* health cures and the future of medicine

★ ★ ★ ★ ★ ★ ★ ★ ★ ★ ★ ★ ★ ★ ★ ★ ★

Part Seven:

The Return of the Past
The Changing
Tides of History   232

* chance events and turning points
* battles won and lost by chance
* assassinations, coups and uprisings
* fate and the line of succession
* chance discoveries
* the mysterious past and the chances of lost cultures
* chance reprieves and sudden fame
* quirks of history

Part Eight:

Through a Glass Darkly
Abracadabra and
the Hand of Destiny 266

* numerology and your fate
* prophecies and predictions
* divine intervention
* faith healing
* miracles and revelations
* fortune, luck and spells
* superstitions
* palmistry and astrology

"This is the plain truth:
every one ought to keep a sharp eye
for the main chance."

PLAUTUS (C. 200 B.C.)

Robert L. Ripley

Introduction

The Chance of a Lifetime

"A chance may win that by mischance was lost."

ROBERT SOUTHWELL (C. 1590)

There is hope. No matter how hard the times, how great the inconsistencies of life, how hard the daily grind of sheer survival, there is hope.

For many of us, it may be that one joker in the deck called Chance. Maybe, just maybe, there is more to life than blind acceptance of fate. Chance can be guised as luck, fortune, kismet or coincidence, call it what you like. But Chance is a reality that may come to us out of the blue, and that may or may not improve our lot, but which at the very least changes things for us. We've all heard about chance events affecting other people's lives, people who've won lotteries, received good raises, come into money, or been blessed with unexpected opportunity. And there's every reason to believe that it just might happen to us too. Certainly in our own lives we can think of things that had they gone a different way would have altered our circumstances. And come to think of it, perhaps most of what we do revolves around *Chance.*

We all yearn for security, for some degree of certainty and peace of mind. But as we are forced to admit, especially in these turbulent times, the future is most uncertain. We are all subject to changes from every quarter, whether it be our job, our marriage, our place of residence, government policies, our friends or even our own attitudes and values.

Despite our interminable efforts to control change—by studying the past, by analyzing the present through statistics, by planning for the future—we may still be sure that chance events remain firmly as the source of almost all change in life.

But it has also been said that there is no such thing as chance. There are certain esoteric disciplines which if practiced with diligence claim that we can anticipate Chance. The theory is that everything that happens to us is of our own making— that we are responsible for our own lives through our conduct, attitudes and perceptions. And that by "living right," as it were, we can control chance and anticipate just what might happen. But for most of us this is either totally impractical or simply absurd. There just isn't time for navel-gazing nor for joining a monastery to ponder the deeper laws of the cosmos. And so, caught up in the daily round of survival, in self-defense we simply dismiss random change as that which we don't understand. And we call it *Chance.*

Perhaps however there is a way to temper "the slings and arrows of outrageous fortune." There is surely a lot of mileage in simply looking on the "bright side of things" and trying to maintain a degree of healthy balance and hope in the face of change. There is also probably some truth in the theory that if we can at least recognize any given chance situation in its beginnings, we can better ride the waves that may follow. One wag has called it "psychic surfing." Because once circumstances have grown to their full consequences they tend to acquire power over us, overwhelm us, and we remain virtually helpless before the full tide and must passively let events take their course. An alert mind, optimism, effort, and a clear and tranquil disposition are perhaps the prime requisites for the smoothest ride through the surf. That may be small comfort in a noisy world moving at such an erratic pace, but it is still true that getting on top of the wave, despite the odds, makes for a more pleasant ride than being drowned by it.

And then, of course, there's always that one big chance that lifts us right to the crest of the wave. Maybe, just maybe ....

But when it's all said and done, perhaps the great Dr. Samuel Johnson summed it up best when he said: "In this state of universal uncertainty, where a thousand dangers hover about us, and none can tell whether the good he pursues is not evil in disguise, or whether the next step will lead him to safety or destruction, nothing can afford any rational tranquility, but the conviction that, however we amuse ourselves with unideal sounds, nothing in reality is governed by chance, but that the universe is under the perpetual superintendence of him who created it; that our being is in the hands of omnipotent goodness, by whom what appears casual to us is directed for ends ultimately kind and merciful; and that nothing can finally hurt him who debars not himself from the divine favour." (*The Rambler*, No. 184, 1751)

The Book of Chance has been compiled with a view to entertain, amuse, and hopefully amaze you with astonishing examples of Chance at work in our lives, There are hundreds of stories here about Chance in all walks and pursuits of life.

Believe it or not, Chance played a vital role in the early beginnings of the man who in essence influenced and made possible this very book. That man was Robert L. Ripley, a small-town boy born in 1893 in Santa Rosa, California.

On December 19, 1918, the sports editor at the old New York *Globe* was anxiously calling for the daily cartoon, and sports artist Robert Ripley was desperate for an idea. The day had been dull for sports, and current news had not provided any subject either.

After looking at his blank drawing paper for some time, Ripley turned finally to his files and a scrapbook of newsclippings he had been putting together for his own amusement. This contained an assortment of sports oddities and records, and he began to draw his cartoon from them. Studying the completed piece of several sports oddities which he had entitled "Champs or Chumps," Ripley was not at all sure that he had produced a good day's work. On impulse he scratched out the original heading, wrote in its place "Believe It or Not!", and put it on the sports editor's desk. Dismissing the drawing from his mind he then went across the street

for a cup of coffee, little realizing that an idea had been conceived that would eventually capture the imagination of millions of readers and make the words "Believe It or Not!" a household phrase the world over.

The following day, the first Believe It or Not! cartoon appeared in the *Globe* and, much to the artist's amazement, drew considerable comment both from readers and from his fellow newspapermen. The editor requested him to draw a similar cartoon as soon as material was available and the idea crystallized. At first it appeared weekly, then twice a week, and finally, every day. Other papers requested permission to use it, and the feature soon became a worldwide institution, appearing in more than 302 newspapers in 38 countries and 17 languages with an incredible readership of 80,000,000!

In his quest for odd and unusual material to illustrate his cartoons, Ripley, sponsored by the great newspaper tycoon William Randolph Hearst and King Features Syndicate, eventually visited 198 countries. One day Ripley might be found posing with a 9-year-old child-mother in Java, or perhaps interviewing a Papan cannibal chief, or looking at the Egyptian pyramids or the Great Wall of China. He was in fact more often in China than in his New York offices. His passion for travel reached such a point that he actually felt he was in a "rut" when he stayed in one place for more than a few months.

His ambition to visit every country in the world to search for strange and fascinating things caused the duke of Windsor to dub him "The Modern Marco Polo."

Ripley was the first artist to send a cartoon by radio—from London to the New York *Tribune*. He also broadcast the first radio program from mid-ocean to a nationwide network in 1931, and in 1934 he was the first to broadcast to various nations in the world simultaneously, assisted by a corps of linguists who translated his message into various languages.

At Chicago's Century of Progress Exposition in 1933, Ripley introduced the first of his famous museums, or "odditoriums," where he exhibited some of the outstanding curiousities which he had amassed from his world travels. By the time of his death in 1949, Robert Ripley had left an astounding legacy. And it was all started by a chance idea that clicked. Robert L. Ripley had come a long, long way from his humble beginnings in Santa Rosa.

Ripley was once asked whether he would ever run out of material for his daily strip. His answer was that it was impossible to run dry on astonishing facts about our world—that as a source of amazing stories it was a bottomless pit. And that is perhaps even more the case now in our rapidly changing world than in Robert Ripley's day. *The Book of Chance* is the first book in Ripley's new publishing program which will offer facts and stories about our astonishing world woven around themes vital to our daily lives. We, the editors, writers and researchers at Ripley Books would like to hear from you the reader. We would like to hear your stories. If you have had an experience in your life, or have heard about, or read of, events that test credibility but which are positively verifiable, send them along to us.

We now offer you one of the most amazing stories about chance we came across. Consider the following:

The Most Important Vote Ever Cast.

Frederick August Conrad Muhlenberg was a member of the Continental Congress of the 13 American colonies, and subsequently Speaker of the First House of Representatives in 1795. He was born in America, the son of a German immigrant, and was educated in Germany.

In 1774, to make the separation of the colonies from England more emphatic it was proposed in the American Continental Congress that the official language of the new political entity be changed from English to German. Twenty-seven members of the congress voted for, and 27 voted against, this proposal. Frederick Muhlenberg broke the tie by casting a negative vote. And the reason for his dissenting vote was that he thought the antiquated German script "would present unsurmountable difficulties" to a thorough mastering of the language.

If this man had voted yes, the implications for subsequent events in world history would have been staggering, including on which side the U.S. might have entered both world wars.

The story of this most fateful vote in American history is contained in Heinrich Melchior Muhlenberg's "Letter" published in Halle, Germany, in 1887. Believe It or Not!

The following 3 cartoons are probably Robert Ripley's all-time most famous Believe It or Not! items:

It was Ripley who helped to make "The Star-Spangled Banner" the official United States anthem. Congress had repeatedly refused to recognize it as such.

On November 3, 1929, a Believe It or Not cartoon was published with this caption: "AMERICA HAS NO NATIONAL ANTHEM. The United States—being a dry country—has been using without authorization—an old English drinking song ('To Anacreon in Heaven')."

As a result of this cartoon more than five million petitions descended upon Congress. There were letters from people in all walks of life. Invariably they referred in shocked accents to the Believe It or Not cartoon. "Can this be true?", they queried, "and why isn't something done about it?" Something was done. On March 3, 1931, Congress finally passed a resolution making "The Star-Spangled Banner" the official United States anthem.

CLINTON W. BLUME
1400 Ocean Parkway, Brooklyn

LOST A SCRUB BRUSH AT SEA, WHEN ARMY TRANSPORT SANK 500 MILES OFF COAST OF FRANCE, AND IT WAS WASHED ASHORE AT HIS FEET IN BROOKLYN ONE YEAR LATER!

"Rip's" Most Doubted "Believe It or Not"

N-X-211
RYAN
NYP

Spirit of St Louis

LINDBERGH WAS THE 67th MAN TO MAKE A NON-STOP FLIGHT OVER THE ATLANTIC OCEAN!

Soon after that flight, Ripley commemorated the event with this cartoon. However, the American public was outraged: it seemed as if Ripley was determined to destroy an American hero. But when the anger finally subsided, Americans realized that Lindbergh's feat was no less heroic simply because 66 men had preceded him. It was the solitude of the Lone Eagle's flight that made it so impressive: one lapse of attention would have killed Lindbergh. So, without depriving Lindbergh of his honor, Ripley's cartoon at least praised 66 forgotten men, including the crews of the British airship R–34 and the German zeppelin ZR–3 and two British airmen, Alcock and Brown, who successfully crossed the Atlantic in 1919.

The Book
of Chance

PART ONE:

It's Show Time!

Show Biz and the Big Break

Your hands are wet, your tongue thick with excitement. Could this be your big chance?

Thousands of stars in movies, television, Broadway and stage began their careers that way, waiting for the big chance. For some it was the night the star was sick and the understudy rushed on; for others it was a beauty contest where a talent scout tapped them on the shoulder. For many it was a lucky encounter or a sudden change from everyday routine.

Some found success when they were forced to take a chance by accident. Others had the big break thrust upon them, even when they didn't want it.

The odds of anyone, aged 16 or over, becoming a movie or TV star are 300,000 to 1 against, yet the annals of Hollywood are filled with stars who owe, or owed, their stardom to the film capital's version of chance—the big break.

Chances are Rock Hudson would never have made it but for a job as a milkman.

And where would Marion Morrison be today if he hadn't changed his name to John Wayne?

Groucho Marx nearly played the lead in *Gone with the Wind,* but the role went, by chance, to Clark Gable.

Gina Lollobrigida nearly walked away from her big break.

There are stories of superstition and curses, stars who discovered new stars, actors and actresses who turned down Oscar-winning roles. Some television stars took roles simply for the money and were astounded when the wheel of fortune finally clicked.

Some movies were almost never made and others should never have appeared. Chance and fate are intertwined through art, writing and music.

How They Got Their Start

Some of the biggest names in movies and television began their careers in unlikely ways. Story after story is told of the big break that would never have happened without that chance encounter, amazing coincidence or lucky streak. Literally hundreds of stars admit they would still be working in obscurity if it hadn't been for that incredible element of chance.

Come with us now as we meet the lucky famous who turned chance into success.

You Oughta Be in Newspapers

Names and faces in the press started these careers:

* Lauren Bacall, a New York model, made the cover of *Harper's Bazaar*. Director Howard Hawks saw it and signed her to a seven-year contract.

* Brigitte Bardot was featured in all her glory on the cover of *Elle* in Paris and Marc Allegret signed her for her first film.

* Richard Burton saw an ad for the part of a Welsh boy in a London newspaper, answered it and won the role from director Emlyn Williams.

* Ali MacGraw was a cover girl for many top fashion magazines when Hollywood finally signed her. She went on to make more than $100 million for the studios in films like *Love Story* and *Goodbye, Columbus*.

There's Nothing like a Dame

An apprentice milliner in London took a chance and played in an amateur production that turned her into a star.

Edith Evans was noticed by William Poel, founder of the Elizabethan Stage Society, who just happened to be in the audience. He cast her in *Troilus and Cressida* at Covent Garden in 1912 and hatmaker Edith turned into Dame Edith, the toast of the London stage.

Thank You, Thank You

As breaks go, this is one of the biggest. Tommy Lee Jones won his first Broadway acting role only 10 days after arriving in New York. And Tommy Lee had never taken an acting lesson. He went on to star in the movie *Coal Miner's Daughter*.

Great Beauties

It's true. A chance win in a beauty contest can lead to stardom. Consider the following:
* Debbie Reynolds got her start by winning the Miss Burbank contest in 1948. Warner Bros. just happened to be located in that Los Angeles suburb and Debbie got a screen test.
* Clara Bow won a beauty contest run by three magazines and received a movie part. She only had a small bit in her first film, but audience mail was so heavy she was made a star.
* Sophia Loren won a beauty contest at the age of 14 and her mother began working to get her into movies.

Bubbling Under

We all dream about the big break when the star gets sick and we get cast into the limelight. Well, it actually happened to:
* Shirley MacLaine, who was understudy for Carol Haney in the Broadway production of *The Pajama Game* when, you guessed it, Haney got sick and Shirley got her big break.
* June Allyson, who was understudying Betty Hutton in *Panama Hattie* on Broadway when Hutton got ill. Allyson took her place and was spotted by a talent scout and cast in the movie *Best Foot Forward*.

Who Do You Think I Am?

A beautiful girl took director Mario Costa's eye on a street in Rome, but he nearly didn't sign one of filmland's top sex symbols. At first the young art student thought Costa was looking for a pickup. But Costa persevered and finally convinced her that he was a director, later introducing us to Gina Lollobrigida.

She Came in from the Cold

A producer for RKO studios spotted a beautiful blonde demonstrating refrigerators at a Los Angeles trade show. But he waited too long before signing her and Columbia moved in, whisking her away from the freezer. At first, the new sex star chose the name Marilyn, but the studio decided it was too close to Marilyn Monroe, so she settled on Kim Novak.

The Girl Next Door

Was it chance that led Carole Lombard into her backyard? A director from 20th Century-Fox spotted her while he was visiting next door. He liked her looks and signed her on the spot.

Saved by a Whisker

It's not always a dog's life. Michael, the famous Irish terrier star of *Peg O' My Heart* and other productions, was rescued from the Los Angeles pound the day before he was to be put to sleep.

CAROLE LANDIS (1919-1948) HAD SEVERAL SPEAKING ROLES IN MOVIES EARLY IN HER CAREER, BUT WAS RECOGNIZED AS HAVING STAR MATERIAL ONLY AFTER APPEARING IN "ONE MILLION B.C."—A FILM IN WHICH SHE DID NOT SAY A WORD

Heard the One about the Drugstore?

Most people think Lana Turner was discovered in a drugstore, but they're wrong.

She had a bit part in a murder film, and it was here that a chance walk across a stage in a tight sweater got directors jumping. That didn't stop would-be actresses from flocking to the drugstore at Hollywood and Vine, however. But no one got anything except a soda.

And a Pint of Cream

The long-reigning romantic lead in Hollywood was discovered as a milkman. Rock Hudson couldn't find acting work so he drove a milk truck to make ends meet. One of his customers was a talent scout who gasped at her handsome deliveryman and sent him off to a screen test.

One Lump or Two?

A favorite character-actor in the 1940s got into acting because his tea business dried up.

Sydney Greenstreet owned a tea plantation in Ceylon but turned to the stage when it went sour. He had read Shakespeare at night while running his spread and could recite 12,000 lines from memory. He's now remembered for his famous role as the evil Fat Man in *The Maltese Falcon*, with Humphrey Bogart.

'Ere, 'Ere

An almost incomprehensible cockney accent nearly robbed us of one of Britain's biggest stars. Michael Caine worked for years to tame his tongue and finally got his big chance when he played Horatio opposite Christopher Plummer's Hamlet. Caine turned his speech to advantage, however, when his career was established, using it in *The Ipcress File* and *Sleuth*.

A Star Is Born

Barbra Streisand became a big star because she had bad teeth.

Smitten with stage fever after seeing *The Diary of Anne Frank* on Broadway, Barbra pleaded with her mother to let her go to theater school. But the only money saved was for Barbra's bad teeth and one of the world's most popular stars finally persuaded her mom that acting was more important than smiling. Barbra was right.

Here's Looking at You, Kid

A director took a chance on a slim young actor and made him into one of the film world's great tough guys. Humphrey Bogart was always considered too small for gangster roles and was cast as a society man in his early films. When he finally got a sinister part, Bogart (his real name) electrified filmgoers everywhere.

Once More, with Feeling

A chance walk past Marlene Dietrich's restaurant table started the career of one of Hollywood's biggest stars. John Wayne was so good-looking that Dietrich turned to a studio head and said, "Daddy, buy me that." The studio did and Wayne starred in hundreds of films. Oddly enough he played war heroes in many, but never served in World War II. He was judged ineligible for combat because he was the only support for his widowed mother.

WALTER BRENNAN THE FILM ACTOR, DID NOT BECOME SUCCESSFUL UNTIL AS A YOUNG MAN MOST OF HIS TEETH WERE KNOCKED OUT IN A MOVIE FIGHT AND HE FOUND *THAT WITHOUT THEM HE COULD PLAY ROLES AS AN OLD MAN*

This Ford's Fixed

Hollywood's biggest stars all had to wait for chance to smile on them, but Harrison Ford nearly gave up. The star of *Star Wars*, *Raiders of the Lost Ark* and *The Empire Strikes Back* didn't have it so good back in 1967. Times were so tough that he found he couldn't scrape by on the $150 he made from television bit parts. So he went out and borrowed a book on carpentry, bought a toolbox and moonlighted as a Mr. Fixit.

Here's Burt

A chance guest appearance on Johnny Carson's *Tonight Show* gave millions their first look at one of the highest paid stars in the business.

Burt Reynolds just wasn't making it in his early bit film parts, but he showed his true colors on the talk-show circuit. Fans loved his quick wit, devil-may-care patter and Errol Flynn–good looks. Thousands wrote to ask more about gorgeous Burt and the studios noticed, casting him in *The Longest Yard*, *Smokey and the Bandit* and *Hooper* — all blockbusters.

Oh, That Face!

Chance turned a little-known architect into one of Italy's biggest stars.

Marcello Mastroianni spent 10 years thinking he'd eventually finish college and become an architect, but his schooling was interrupted when he was captured by the Nazis during World War II. He escaped and supported himself in Venice by drawing pictures on handkerchiefs in return for food. He was then noticed by Luchino Visconti who offered him a part in *A Streetcar Named Desire*, but Mastroianni never took the offer seriously. Then Fellini cast him in *La Dolce Vita*, which made him a star, because Mastroianni had a face "with no personality in it."

Down from the Top

A top executive at the American Broadcasting Co. took a chance and tried acting.

Telly Savalas decided to play a judge in a TV show when no one could be found with the right magisterial qualities. He went on to many movies, including *The Dirty Dozen*, and scored his big hit with TV's *Kojak*.

That's Life

A cover shot on *Life* showing her in a wet bikini led Hollywood to one of its biggest finds. The girl turned out to be Racquel Welch and *Life*'s loss has become our gain.

Easy as...

One of the biggest names in Hollywood got his start because he could fall off a horse.

Gary Cooper wore his best suit to a tryout for a western movie, but suspicious producers thought the big actor was a dude and made him prove he could ride—and fall off—a horse. He went on to a career that culminated in the classic *High Noon*, but before he made it big, Coop was fired and rehired by the movie bosses seven times.

Play It Again, Walt

His singing at a party made him a star. Walter Pidgeon was discovered by Fred Astaire when he heard him sing, and Fred immediately recommended him to friends in the movie business.

But Paul Newman Was Great

The El Rey Theater in Manteca, California, burst into flames shortly after a showing of *The Towering Inferno*. The fire, of undetermined origin, leveled the theater.

Still a Hero

Jack Nicholson was in the newspapers long before he became a film star. By chance, Nicholson was working as a lifeguard in New Jersey in the mid 1950s when 11 swimmers were carried out into the Atlantic. Nicholson launched one of the boats and rescued five of the swimmers just as they were about to go under. His picture was on the front page of local newspapers, but Nicholson later said of the rescue that he was so sick "I puked my guts out."

From Here to Columbia

A once-famous singer offered to work free in *From Here to Eternity* because he thought he was washed up.

Frank Sinatra got the part of Sergeant Maggio for a small salary and the movie led to a new start as an actor. It also produced a generation of fans who wanted him to sing again. Thirty years later Sinatra is still the biggest in the business in Las Vegas.

And He Couldn't Sing, Either

Famed for his musical extravaganzas and elaborate choreography, Busby Berkeley never took a dancing lesson in his life.

She Loved It

An ad-lib gave us the first talking movie.

Al Jolson was making *The Jazz Singer* when he turned to the actress playing his mother and said, "Did you like that mama? I'd rather please you than anybody I know of." The words sounded so effective to producer Sam Warner that he left them in when the movie opened in 1927. They were the only spoken words in the movie—the rest of the "talking" was songs.

And Now for Something Completely Different

Although they'd like to forget it, some of the most famous movie stars in the world were forced to sing for their paychecks. Among them:

* James Stewart signed a contract with his first studio that allowed them to make him sing.

* Clark Gable actually sang and danced on screen in his early films.

* Lee Marvin and Clint Eastwood both sang in the musical *Paint Your Wagon*.

* Kirk Douglas played a conch ukulele and sang "Mermaid Millie" in Disney's *20,000 Leagues under the Sea*.

* Sean Connery, Hollywood's first James Bond, sang a love song in the cult classic *Darby O'Gill and the Little People*.

It's a Better Climate, Too

A heart attack led a favorite character-actor to Hollywood.

Dan Duryea suffered a stroke while working in the advertising business and moved to California and the movies to escape the rat race. He went on to star in *The Little Foxes* and *The Pride of the Yankees*.

He Kept the Shift Work

Despite appearances in *Going My Way* and *How Green Was My Valley*, a famous actor was unsure movies would ever support him. Barry Fitzgerald kept his British civil servant job for 14 years before he finally decided to stick to movies. Oddly enough, Fitzgerald became famous for playing Catholic priests even though he was a Protestant.

How Do You Want Your Stake?

A classic film scene became almost too real for a famous star. Jean Seberg was playing St. Joan in the climactic stake scene when her clothes caught fire and she nearly got grilled.

I Love You, Archie

Who would have believed that Archibald Leech from Bristol, England, would become the king of Hollywood leading men? Thank heavens Arch changed his name to Cary Grant and got his lucky break.

Superstar or Spy?

Was Hollywood's greatest adventure star a spy for the Nazis? Since his death, controversies have raged surrounding the life of Errol Flynn. Among them was his association with a German doctor who rescued him off New Guinea in 1933 when Flynn was shipwrecked. The actor later used his influence in Washington—especially his friendship with the wife of the president—to help the doctor retain American citizenship during World War II. The doctor was later proven to be a full-time Nazi agent and his association with Flynn had enabled him to spy freely during the war.

But that was only part of Flynn's incredible life—a life ruled by chance, fate and luck.
* He was a descendant of Fletcher Christian and Edward Young, two of the original mutineers on the infamous *Bounty*.
* He believed in reincarnation and thought he had been a pirate in a previous life, a convenient metamorphosis that helped him play roles like Captain Blood.
* Flynn idolized John Barrymore, saw all his films, collected photos of him, used a cigarette holder identical to Barrymore's, and carried a monkey on his shoulder because Barrymore had a pet ape.
* Flynn's first big movie, *Captain Blood*, made him an overnight success and the idol of women around the world. But he got the part because a British actor signed for it and then later backed off because he felt California would be bad for his asthma.
* In 1938 Flynn's yacht was banned from almost every port on the American west coast because he was suspected of signaling Japanese fishing boats and secretly taking films of Pearl Harbor.
* Flynn refused to enlist at the beginning of the war because he was a British citizen. After Pearl Harbor he still didn't enlist, for it was discovered he had tuberculosis. When he died, he also had sinusitis, emphysema, gonorrhea, chronic irritation of the urethra and a heart murmur.

MACK SENNETT

PRODUCER OF THE KEYSTONE COPS AND OTHER FAMOUS FILMS, BEGAN HIS CAREER *PLAYING THE HIND LEGS OF A STAGE HORSE*

Big Mistakes

Chance works in mysterious ways. Just consider these people who refused to take a chance:

* George Raft turned down the roles that made Humphrey Bogart famous: *Dead End*, *High Sierra*, *The Maltese Falcon* and *Casablanca*. Bogart ended up as an immortal star and Raft finished his days running a gambling club.

* Producer Harry Cohn of Columbia let Marilyn Monroe's contract lapse in 1949 because he decided she lacked "star quality."

* 11 actors turned down the role of the crooked insurance salesman in *Double Indemnity*. The part finally went to Fred MacMurray and became a film classic.

* Doris Day turned down the role of Mrs. Robinson in *The Graduate*, one of the biggest money-makers of the 1970s.

* Lee Strasberg, head of the famed Actors Studio, once told Robert Blake he could never learn to act. Blake went on to star in the popular American TV show *Baretta* and was voted outstanding actor in a dramatic series in 1975 by the U.S. Academy of TV Arts and Sciences.

RICHARD HAYDN
WHO BECAME A HIGHLY SUCCESSFUL
DIRECTOR AND ACTOR, STARTED
HIS SHOW BUSINESS CAREER
IMITATING FISH

Meet the Producers

They had to start somewhere and here are some of the most remarkable beginnings in the movies:

* Harry "King" Cohn was a school dropout from New York who went west to plug songs. He eventually founded Columbia studios where his difficulties with the English language became famous. His own executives used to bet him he couldn't spell the studio's name and Cohn usually lost.

* Adolph Zukor, head of the early Vitagraph studios and one of Hollywood's biggest producers, started as an immigrant from Hun-

gary, sweeping out a drugstore.

* Mack Sennett, originator of the Keystone Kops, began as a boilermaker.

* Lewis J. Selznick, father of David and creator of the star system, went broke hiring opera stars for silent movies.

* The head of Universal studios, Carl Laemmle, was a German bookkeeper from Oshkosh, Wisconsin.

* Marcus Loew, founder of Loew's theater chain, was a furrier from Manhattan.

* Louis B. Mayer began as a junk dealer from Minsk.

Big Bombs

Hollywood is not full of happy endings. Many surefire films turned out to be stinkers at the box office despite top direction, millions of dollars and big-name stars. Chance was cruel to:

* *Intolerance*, a 4½-hour epic made by D. W. Griffith to answer charges of racism in his previous opus, *Birth of a Nation*. It cost a million dollars back in 1915 and died when the United States entered the Great War.

* *Freaks* was made by *Dracula* director Tod Browning, who thought the taste of the movie public ran to the grotesque. All the freaks were real and so was the public disgust.

* *The Great Gatsby*, a multimillion-dollar remake of the classic 1925 novel, was advertised as having everything—big stars, huge sets, wonderful costumes and top direction. It had everything except an audience.

* *Heaven's Gate* was budgeted at $30 million by director Michael Cimino but it grew to $60 million. The studios backed Cimino because of the success of his *Deer Hunter*, but the movie, four hours long, put everyone to sleep except the critics who savaged it. An attempt to re-edit it and show it in second-run houses didn't work either.

* *The Exorcist, Part II* was ballyhooed to make Part I look like a stroll in the country. But Part II, despite Richard Burton, was unforgettable only for a view of an African village through the eye of a fly. The movie scared nobody but its backers.

* *Lost Horizon*, a musical remake of the Utopian tale first shot in the 1930s, crashed like the downed plane in the story.

Double Chance

Everyone knows Grace Kelly got her big chance when she ran off to marry Prince Rainier of Monaco. But did you know Grace's big break also gave Elizabeth Taylor her big chance? Famous as a child actress, Liz's career was stuttering when she got Kelly's part in *Cat on a Hot Tin Roof* and she used it to become one of the biggest stars in the world.

Who Was That Tramp I Saw You with Last Night?

BARRY FITZGERALD

THE STAGE AND SCREEN ACTOR BEGAN HIS CAREER WHEN AS A VISITOR BACKSTAGE AT DUBLIN'S ABBEY THEATER HE WAS *ACCIDENTALLY SHOVED ONSTAGE DURING A MOB SCENE*

Hollywood's most famous character in the silent era of films started with some chance rummaging in a wardrobe chest.

Charlie Chaplin stuck a piece of black crepe under his nose, borrowed Fatty Arbuckle's hat and pants, Chester Conklin's coat and added his own cane. Mack Sennett noticed him clowning around and told him to use the character in the 1914 picture *Kid Auto Races at Venice.* Chaplin called his creation the Little Tramp.

Hollywood Quirks

Besides the big break, the chance discovery and the lucky encounter, Hollywood, vaudeville and Broadway are full of wheel-of-fortune stories. Imagine what would have happened to these stars if the wheel hadn't clicked their way.

School of Hard Knocks

One of Hollywood's favorite comedians got his name because of a fall. Buster Keaton tumbled down the stairs when he was a tot and Houdini saw him. The great magician said "that's a real buster," and the name stuck. Keaton made a career of falls in comic bits. When he died X-rays showed he had broken every bone in his body at one time or another. But Keaton never wore a cast and once reported to work with a broken back!

All in the Family

What are your chances in Hollywood if you're the son or daughter of a famous star? Pretty good, to judge by the following:
* Jeff and Beau Bridges, sons of Lloyd
* Alan Alda, son of Robert
* Shaun and David Cassidy, sons of Jack Cassidy and Shirley Jones
* Alan Ladd, Jr., producer
* Jane and Peter Fonda, progeny of Henry
* Andrew Stevens, son of Stella
* Liza Minnelli, daughter of Judy Garland
* David and Keith Carradine, sons of John
* Nancy and Frank Sinatra, Jr.
* Jamie Leigh Curtis, daughter of Janet Leigh and Tony Curtis
* Carrie Fisher, daughter of Debbie Reynolds and Eddie Fisher
* Michael Douglas, son of Kirk

A True Outlaw

The man who became a star playing bandits was a real-life desperado. Rudolph Valentino, star of *The Sheik* and *Blood and Sand*, fled to Hollywood to escape the irate husband of one of the women he was seducing. Critics still charge that Valentino simply played himself in movies....

Jayne's Curse

A bizarre curse may have led to the death of one of America's top pinup actresses. Call it chance or fate, but the story of buxom Jayne Mansfield is one of the strangest in Hollywood. And it is still going on, despite the fact she was killed in 1966. We bring you this tale:

* Jayne's world was going to pieces in the mid 1960s, shortly after she'd bought a huge pink million-dollar mansion and installed a heart-shaped bed. She was going through her third divorce and was being blackmailed by her lawyer, who threatened to expose her as a high priestess in Anton Lavey's Church of Satan.

* Lavey put a death curse on Jayne's lawyer, and soon after the lawyer went for a ride with Jayne and her friend Ron Harrison. Harrison drove straight into a truck and he and the lawyer were killed. Jayne was decapitated and her head, blonde tresses flowing, ended on the hood of the car.

* The curse didn't end there. The new owner of Jayne's house watched in horror as his son drove Jayne's pink car through the front gates and was killed.

* The second owner was Mama Cass Elliott of the singing group the Mamas and the Papas. She choked to death on a sandwich in London while the house was being renovated.

* The third owner became possessed and convinced herself she was a reincarnation of Jayne. She wore the actress's old clothes, had a breast enlargement operation and had her hair bleached and styled like Jayne's.

* The house was then sold to Beatle drummer Ringo Starr, who quickly resold it after it mysteriously returned to its original pink color—no matter how many times he painted it white.

* The fifth owner was singer Engelbert Humperdinck. He had the house blessed by a Catholic priest before he moved in, but found a heart-shaped crater in the ground after an earthquake.

Chance or fate? The story didn't end with the house. Mickey Hargitay, Jayne's second husband, had a bad accident driving through the gates of the property. Victor Hudson, Jayne's road manager, died shortly after her death. And finally, the writer of Jayne's biography found blood on the pages describing her death.

Call from the Grave

Jean Harlow's husband killed her—long after his own death. During one of the many quarrels in the course of their marriage he struck her and damaged her kidney. She died of kidney failure, but only after he had been dead for three years.

Death Car

One of the strangest legends in Hollywood surrounds the car that killed James Dean in the 1950s. Dean himself called the car a "little bastard" but we find it was something more than that:

* Dean was killed in the car in a head-on collision at 80 miles an hour, but an investigation of the crash showed Dean, an expert driver, apparently made no effort to avoid the fatal smash.

* Dean's reputation as an actor in *East of Eden*, *Rebel without a Cause* and *Giant* turned the Death Car into a trophy. But when a friend first tried to sell it, the car fell on a garage mechanic and broke his legs.

* Two doctors bought parts from the car to reuse in race cars. One was killed and the other seriously injured.

* Fans who arrived looking for souvenir pieces of the car were hurt trying to remove them.

* The only two undamaged tires were sold to a man who was sent to hospital after they both mysteriously blew at the same time. Later inspection failed to find a fault in either tire.

* The California Highway Patrol planned to use the remains of the car in an auto show, but the night before the show opened, fire destroyed every vehicle. Every vehicle, that is, except Dean's car, which escaped without a scratch.

* The car was later loaded on a truck and sent to Salinas, Dean's destination on the day of the crash. The truck driver lost control and was thrown out of his cab and crushed. But the Dean car rolled safely off the truck.

* In 1959, when the car was on display in New Orleans, it suddenly broke into 11

pieces even though it had been carefully welded together.

* Strangest of all was the loan of the car to the Florida police for a safety display in Miami. After it was crated and put on a truck, it disappeared and has never been seen since.

This Old General Died

An Academy Award role was lost because the star felt too good. Rod Steiger turned down the lead in *Patton* because he felt too well to take on a role that glorified war. The part, and an Academy Award, went to George C. Scott.

As Time Goes By

The piano player in *Casablanca* responsible for the famous tune couldn't play a note. Dooley Wilson was a singing drummer. And the line everyone remembers, "Play it again, Sam," was never said in the movie. So there.

Ronnie Reagan's Lucky Strike

HE SAVED **77** PERSONS FROM DROWNING IN HIS **7** YEARS AS A LIFE-GUARD AT A RESORT NEAR DIXON, ILLINOIS--
HE CELEBRATED HIS **70**th BIRTHDAY **17** DAYS AFTER HIS INAUGURATION--
WOUNDED BY A WOULD-BE ASSASSIN ON HIS **70**th DAY IN OFFICE, THE BULLET RICOCHETED OFF HIS **7**th RIB--
HE MADE HIS FILM DEBUT IN 193**7** AND BECAME PRES. OF THE SCREEN ACTORS' GUILD IN 194**7**--
HE BEGAN HIS TERM AS GOVERNOR OF CALIFORNIA IN 196**7** AND WAS REELECTED IN 19**70**--
HIS FORMAL ACCEPTANCE SPEECH OF THE REPUBLICAN PRESIDENTIAL NOMINATION WAS MADE ON **7-17**, 1980--
IF HE SERVES TWO FULL TERMS IN THE WHITE HOUSE HE'LL BE **77** WHEN HE LEAVES--

I'll Never Forget What's His Name

Would chance have given these stars a break if they'd stuck by their original names? We doubt it:
* Eugene Orowitz (Michael Landon)
* Harriette Lake (Ann Sothern)
* Bernie Schwartz (Tony Curtis)
* Lucille Le Sueur (Joan Crawford)
* Harlean Carpenter (Jean Harlow)
* Edythe Marrener (Susan Hayward)
* Norma Jean Baker (Marilyn Monroe)
* Marion Morrison (John Wayne)
* Frances Gumm (Judy Garland)
* Joe Yule (Mickey Rooney)
* Phyllis Isley (Jennifer Jones)
* Margarita Cansino (Rita Hayworth)
* Gladys Smith (Mary Pickford)

Horror Stories

Chance, luck, superstition and fate. Horror stories have it all. But the tale behind the screen story is sometimes even odder:
* *Frankenstein* grossed more than $12 million during the Depression but cost only $275,000 to shoot.
* Boris Karloff, the world's most famous Frankenstein, started in the business telling children's stories on BBC radio.
* Lon Chaney, the original Phantom of the Opera and one of Hollywood's great horror stars, started in local theater playing clowns.
* Bela Lugosi only made $3,500 from the movie *Dracula*, but acquired a role for life. He was buried in his vampire costume.
* Judy Garland decided to appear in *Valley of the Dolls* even though she knew her character—a hopeless drug addict—was based on herself. Garland died of a drug overdose before the film went into production.

Call Me in the Morning

Lew Ayres, the first Dr. Kildare, was a dropout from medical school.

The Wardrobe Chest

Take a chance, open the chest and we find:
* David Niven wore the same old trench coat in at least one scene of every movie he ever made. It was his own coat and Niven believed it brought him luck.
* Carroll Baker started a nationwide fad in the movie *Baby Doll*. The name of her charter was Baby Doll and—you guessed it—she wore baby-doll pajamas. The style is still around.
* Gina Lollobrigida and Elizabeth Taylor once ran into each other in Moscow. They were wearing the same dress.
* Sophia Loren made both the Best-Dressed and Worst-Dressed Women lists in 1972.

The Longest Chance of All

No movie has shown the magic, misfortune and mystery of chance like *Gone with the Wind*. It's a miracle it ever got made. The female lead wasn't found until the film was being shot, the two big stars hated each other, the cast and crew threatened to walk out many times. Here are some of the stories behind Hollywood's greatest film:

* Producer David O. Selznick had to be talked into paying $50,000 for the rights to Margaret Mitchell's epic novel. He only bought the book after someone else threatened to grab it first.

* Katharine Hepburn figured she was perfect for the part of Scarlett O'Hara but refused to take a film test, saying Selznick should know she was right for the part. He turned her down cold and also rejected Lucille Ball, Susan Hayward, Paulette Goddard, Bette Davis, Joan Crawford, Tallulah Bankhead and Ann Sheridan.

* So desperate was Selznick to find the perfect Scarlett that he decided to hold an America-wide seach to cast an unknown in the part. A deluge of applications followed, but Selznick had to go into production without his female lead.

* Selznick was standing on a balcony directing the filming of the burning of Atlanta when his brother, Myron, showed up with Vivien Leigh and told David he had found their Scarlett. Myron was right, but Leigh accepted the part reluctantly because she thought Americans would not accept an English girl in the part of a Southern belle.

* Mitchell wanted to cast Groucho Marx in the role of Rhett Butler, but cooler heads prevailed and she was talked out of it. The studio wanted Gary Cooper or Errol Flynn, but neither was available so it settled on Clark Gable. Although he didn't want the part, Gable took it to raise money for his marriage to Carole Lombard.

* Leslie Howard, the famous British star, didn't want to play Ashley Wilkes because he was tired of weak-kneed characters. He agreed only after the studio said he could make *Intermezzo* next with Ingrid Bergman.

* Joan Fontaine was invited to read for Melanie but she sent the part to her sister, Olivia De Havilland.

* More than 10 writers, including F. Scott Fitzgerald, were involved in the final script but Selznick didn't like any of it. He kept going back to the novel to see if he could find better dialogue.

* Directors didn't fare much better. The film started with George Cukor but Gable didn't like him. Selznick then hired Victor Fleming but Leigh and De Havilland didn't like him. Fleming had a nervous breakdown and Sam Wood was hired. Sidney Franklin, William Wellman and Reeves Eason did some dramatic scenes and the battles. Jack Cosgrove did special effects and William Cameron Menzies oversaw many scenes.

* Leigh didn't like the love scenes because she said Gable had bad breath. Gable didn't like her because he thought she was prissy.

* The famous line, "Frankly, my dear, I don't give a damn," was almost cut because of censorship rules. It was finally allowed because censors felt it was essential to the story, but the studio paid a $5,000 fine.

* The film finally opened after spending more than double its $2-million budget. Selznick insisted on noiseless programs at the premiere because he didn't want any dialogue to be drowned out by rustling paper.

* The film won eight Academy Awards—every major one except best actor. People felt Gable was merely playing himself.

* Selznick spent the rest of his life trying to top *Gone with the Wind* but everything else was a failure. He went bankrupt with *A Farewell to Arms*, one of the all-time box-office losers.

GONE WITH THE WIND
THE MONUMENTAL MOVIE MADE IN 1939, RESULTED IN THE SHOOTING OF 449,512 FEET OF FILM-- OF WHICH ONLY ABOUT 20,300 FEET APPEAR IN THE FINAL PICTURE

Three Legends

Call it chance or coincidence, but *The Wizard of Oz*, Hollywood's first color picture and a classic movie, has some incredible stories behind it:

* L. Frank Baum, author of the book, came up with the name Oz from the letters on his bottom file drawer, O-Z.
* Judy Garland got the part of Dorothy even though the studio thought she was too old at 17. And Judy was paid less than anyone in the film except for the dog, Toto.
* The film was made because Sam Goldwyn wanted a children's story as a pilot for the Technicolor process.
* Ray Bolger, a song and dance man, was originally cast as the Tin Man, but switched to the Scarecrow when he persuaded Buddy Ebsen to change. Ebsen never made the film, however, because the paint used to turn him into tin affected his lungs. Jack Haley replaced him as the Tin Man.
* W. C. Fields was supposed to play the Wizard, but wanted too much money. Ergo, Frank Morgan.
* Gale Sondergaard was originally cast as the Wicked Witch of the West, but it was decided that she was too glamorous. Twenty-year-old Margaret Hamilton was hired instead and went on to enjoy a 30-year career playing witches and baddies.

Wild West Show

By one of the oddest chances of all, the star of Hollywood's first western was Buffalo Bill Cody.

Unlikely Hero

Can you identify the following actor, one of the most improbable stars of all time?
* He was expelled repeatedly from school.
* He broke into movies by mastering a vicious horse that almost killed him.
* He's the only movie star who made a million a year despite three jail sentences.
* He refused to shave his chest for certain scenes, and he developed a pot belly because he hated beefcake photos.
* He made 84 films but Katharine Hepburn said he couldn't act.
* He received an award in 1949 for being Hollywood's sour apple and accepted it with ease, adding it would go with his elevation to the Worst-Dressed List and nomination to the roll of Hollywood's 10 most-undesirable guests.
* He wrote a song for his own hit movie, *Thunder Road*, sang it and had a hit record.
* A brawler and a drinker, he was never late for work and had always memorized his lines.

Give up? Meet Robert Mitchum.

Scream and Scream Again

One of the most famous sounds in Hollywood was made up of
* a man yelling,
* a hyena howling,
* a soprano singing,
* a dog growling and
* a soft violin playing.

They all combined to make the famous Tarzan scream (by chance, the hyena's growl was recorded backwards). The man doing the yelling became so adept at it that he could duplicate it anywhere. And that's why Johnny Weissmuller is still remembered as the most famous Tarzan of all.

You Ought to Be in Pictures

What are the chances of making it in Hollywood? In the last few years, 75 percent of all actors made less than $2,500 annually and only 3 percent made more than $25,000.

Rest in Peace

Hollywood is full of stories of stars who seem to come back and haunt their favorite spots after death. Is it chance? Judge for yourself:

* Errol Flynn's yacht was known worldwide for wild parties, drinking bouts and, in one case, a paternity suit. After the star of *Captain Blood* and *Robin Hood* died, the yacht was moored off the French Riviera. But neighbors swear Flynn appeared on deck one moonless night and the lights and sounds of a party were to be seen and heard on board.

* John Wayne's yacht was sold after his death but the lanterns were missing. The owner bought new lamps, but later found the originals hidden away. Oddly enough, the new lanterns were the same as Wayne's— the Duke had removed them to keep from bumping his head.

* John Barrymore joked with Rudy Vallee on a radio show that it would be his "farewell appearance." He then collapsed and died the following week.

* Circus stars Emmett Kelly and Karl Wallenda both headlined the Ringling Bros. and Barnum and Bailey show in New York. Both died on the day the circus opened in consecutive years—aerialist Wallenda on March 22, 1978, and clown Kelly on March 28, 1979.

A Passing Fad?

The man credited with inventing the motion picture system never made a dime. Thomas Edison didn't bother to patent his projection systems because he thought they had no future.

By the time he died, Edison had invented more than 1,000 other devices and processes, including the light bulb. But the greatest inventor in America lived the life of a crackpot, working beside rats in his filthy lab, wearing clothes for days because he thought taking them off induced insomnia, and starving himself because he thought food was poison.

That Classic Profile

Accidents—and chance—had a big effect on the career of Marlon Brando.

Brando's profile was thought too perfect until he broke his nose in a fight. His career, broken nose and all, took off after he played toughs in *On the Waterfront* and *The Wild One*.

BEN TURPIN

THE FAMED CROSS-EYED SCREEN COMEDIAN
WAS DISCOVERED WHILE WORKING
IN A FILM STUDIO AS A JANITOR

Remember the Scene When…?

Chance can ruin the best-made film. Despite continuity chiefs and editors, some great bloopers have appeared:

* In *Carmen Jones*, the camera crew is reflected in windows as it tracks star Dorothy Dandridge down a street.

* In *The Wrong Box*, a British comedy set in Victorian England, all the houses have TV aerials.

* In *Decameron Nights*, star Louis Jourdan stands at the bridge of a 14th-century sailing ship while a truck roars around a hill in the background.

* In *Scipio Africanus*, a million-dollar opus made during Mussolini's era, hundreds of soldiers salute Caesar wearing wristwatches. And the Seven Hills of Rome are covered with electricity poles.

All Wet

Her famous giggle was primed by buckets of water. Goldie Hawn got her start on U.S. TV's *Laugh-In* because the director liked her air of befuddlement. Her giggle was best when it was totally spontaneous so the crew started dumping water on her head—one of the best-known bits in American television.

Hard Luck, Charlie

More people saw a wedding on the long-running British TV series *Coronation Street* than the real-life wedding of Lady Di and Prince Charles. *Coronation Street* drew 15 million and the real thing only 13 million.

Angels Have Feelings Too

Chance and an ironclad TV contract bumped her out of an Oscar role. Kate Jackson, one of the original Charlie's Angels, lost the role of the wife in *Kramer vs Kramer* because she couldn't get out of the show. Meryl Streep won an Oscar and Kate has since quit the Angels. But it was too little, too late.

Lost in Space

A veteran character-actor took a chance on a TV series he thought was dumb because he needed the money. Ray Walston, a star of *Damn Yankees* and *South Pacific*, was dumbfounded when *My Favorite Martian* became a huge hit. But his premonition about the show was right. It typecast him for life and he had difficulty landing serious roles.

C'est la Vie

French newspapers in 1978 touted an interview show as the best bet in prime time that evening. But an opinion poll taken while the show was on revealed it had *no viewers* at all.

Blockbuster No. 1

More than 80 percent of the population of Italy saw the eight-hour TV film *Jesus of Nazareth* by Franco Zeffirelli. It was later shown around the world to 600 million.

Blockbuster No. 2

Roots, by Alex Haley, drew the largest single-showing audience in television history, encompassing most of the viewing public of America. Ironically, Haley was later sued by an author who claimed the story had been lifted from his book.

Adjust Your Set

Chances are these were the worst movies ever shown on TV. At least that's what the critics say:
* *Zebra in the Kitchen*. Just what it says.
* *Zombies on Broadway*. Bela Lugosi is a victim of voodoo PR men.
* *Canadian Mounties vs The Atomic Invaders*. The Mounties win, you lose.
* *Merrily We Go to Hell*. Sylvia Sidney marries beneath her station…and ours.
* *Two Little Bears*. A father comes home at night to find his children turning into bears, proving once again that you can never tell what the kids will get up to next.
* *Hercules against the Moon Men*. Battle of the centuries.
* *Spy in Your Eye*. Dana Andrews gets a camera in his cornea.

Boob Tube Mania

The most popular entertainment medium in the world, television is so powerful that the chances are 25 to 1 that, in any given house with a set, someone will be watching. Yet many of the most beloved shows and stars were last-minute replacements or lucky combinations that clicked with the public.

Literally hundreds of TV pilots are launched each year in the U.S. in the hope of cashing in on the television jackpot, but only one in 40 ever becomes a series.

Diving for Pearl

A chance appearance on *What's My Line?* at a time when her career was slipping gave a famous singer and actress her big break. Pearl Bailey was a guest and David Merrick was a panelist. So captivated was Merrick that he offered Pearl the lead in *Hello, Dolly!* on Broadway. The all-black production gave the show a new lease on life and helped Pearl's career.

Father Who

The television actor famous for playing Dr. Who in the BBC series nearly ended up a monk. Tom Baker spent seven years in a monastery before finally deciding to kick the habit.

MAUREEN O'SULLIVAN and MIA FARROW
MOTHER AND DAUGHTER STARRED SIMULTANEOUSLY IN HIT BROADWAY PLAYS AND EACH IS *THE MOTHER OF 7 CHILDREN*

BOBBY CLARK
THE AMERICAN STAGE COMEDIAN KNOWN FOR HIS PAINTED "SPECTACLES" STUMBLED ON THAT IDEA IN 1906 BECAUSE JUST BEFORE A PERFORMANCE *HE MISLAID HIS GLASSES*

Did You Know?

Glenn Ford was born Gwyllen Ford, but renamed himself after his father's Canadian birthplace—a little Quebec town called Glenford.

And Humphrey Bogart was not the first choice to play in any of his major films. Hollywood wanted Edward G. Robinson for *The Petrified Forest*, Ronald Reagan for *Casablanca*, and Gregory Peck for *The African Queen*.

But Bogey hung in and won the parts.

What's His Line?

A mystery guest on *What's My Line?* was a Kentucky colonel who sold fried chicken. The old gent charmed the audience and the panel, but everyone forgot him until his first franchise came out.

Meet Colonel Sanders.

It's All Chance

The King of music believed his entire success was due to blind luck.

Elvis Presley felt the death of his twin brother at birth meant he had inherited his brother's talents by chance. And Presley was undeterred when Arthur Godfrey's *Talent Scouts* turned him down in 1955. He was convinced he would be a success and made his first hit record a year later.

Chance in the Music Business

For many musicians, a spinning record is the wheel of fortune. Even though careers seldom last beyond five years, the recording business is filled with stories of chance, luck and breaks.

Statistics show that only one quarter of the records released in any given year make money and only 17 new artists a year ever break into the Top 40. Yet many take a chance, cut a record and become overnight hits.

Country Roads

She got started because she looked like Hayley Mills.

Olivia Newton-John was only 12 when she won a movie contest to find the girl who most looked like Mills. The prize was a part in *Green Pastures*, her first role, and it started her on the way to a string of hits.

The Sound of Music

A young singer was warbling away in the Catskill Mountains when he caught the ear of Eddie Cantor. He liked young Eddie Fisher so much that he asked him to join his troupe and Fisher went on to become a solo artist and movie star.

Ahem...Ahem

She was toiling away as a member of the chorus when the featured singer at a radio station got laryngitis. Could she fill in? Of course—and the world met Patti Page.

Your Floor, Sir

An usher at the Paramount Theater stopped his elevator between floors to sing for Perry Como. He liked what he heard and Vic Damone eventually won an audition on Arthur Godfrey's show and a club date through Milton Berle. He became a headliner in the early '50s and '60s, but his luck ran out in 1971 and he filed for bankruptcy.

Mr. White Shoes

A fresh-faced young singer forgot all his shoes except for a pair of white bucks when he auditioned for Ted Mack and Arthur Godfrey. But those shoes became a fad and the trademark of who else but Pat Boone.

Okay Dad

She wasn't big on the song, but her father insisted she record "Who's Sorry Now?" back in 1958. Connie Francis doesn't regret it today. The song was her first big hit.

Not So Easy

Chances are you never thought writing music was easy. But these stories defy the odds:

* Siri Nome, a Norwegian pianist, composed a piece called "In C" in which the C note is played nonstop for 26 minutes. When the houselights came on after his performance, the hall was empty.

* Alan Jay Lerner took two weeks to write the last line of "Wouldn't It Be Loverly" for *My Fair Lady.* The line is "Loverly, loverly, loverly, loverly."

One-Hit Wonders

Chance can make—or break—any group or artist. Meet the people who each had a gold record, but never hit it big again. Who can forget:

* "Wiggle, Wiggle" by the Accents, a song celebrating the sack dress in 1958. It lasted about as long as the dress.

* "Bobby's Girl" by Marcie Blaine. She was big in 1962, but a year later was gone from the scene.

* "Cry" by the Bonnie Sisters. They gave up their nursing jobs in 1956 and their patients probably missed them.

* "Funny How Time Slips Away" by Jimmy Elledge (1961). Funny…

* "Rockin' in the Jungle" was recorded in 1959 by the Eternals. They weren't.

* "Susie Darling" was recorded in 1958 by Robin Luke about his 5-year-old sister. He went on to become a college professor.

Que Sera, Sera

Doris Day started her career as a dancer. She turned to singing only after a broken leg put her in hospital for a month.

The March King

The composer of the "Stars and Stripes Forever" received only $90 for all rights to the music. But John Philip Sousa nearly didn't write it. His father persuaded him to enlist in the U.S. Marine Band because he was afraid his son planned to become a circus musician.

Ignace Pleyel (1737–1831), celebrated French composer, was sentenced to death in 1792 during the French Revolution. With the guillotine staring him in the face, Pleyel spent a furious 8 days and nights composing a musical play glorifying the Revolution, and was pardoned and released after the play's successful performance. Believe It or Not!

He Could Still Be in the Mines

One of the world's most popular entertainers turned to singing because he had run out of jobs.

Dean Martin had been a coal miner, a boxer, a gas station attendant and a millhand. In 1946 he decided to sing and landed a club date. He then met a guy named Jerry Lewis and the rest is history.

Remember When?

A French chef injured in a kitchen explosion lost his memory in 1772 and began composing music. Louis Maria Messigny (1742–1832) wrote 31 operas in the next 30 years before chance stepped in again. He was hurt at the age of 60 when a stage collapsed, causing him to regain his memory. He knew how to cook again, but never wrote or read another line of music.

Omens at Altamount

Sometimes too much is left to chance. Take the famous Rolling Stones rock concert at Altamount, California, in 1969, where a fan was stabbed to death. Here are some of the signs that weren't noted:

* A young woman predicted in *Ramparts* magazine that the concert could be a disaster because the sun, Venus and Mercury were all in Sagittarius.
* Jesus freaks passed out leaflets saying the Stones were in league with the Devil.
* During the early part of the concert, Hell's Angels security guards beat up several mem-

bers of the audience and one member of the Jefferson Airplane. But the concert went on.
* After a 1½-hour delay, Mick Jagger and the Stones appeared. Jagger said "something always happens when we sing this song" and launched into "Sympathy for the Devil." Someone pulled a gun and was stabbed and beaten to death while Jagger sang.
* Altamount finally ended with one knifed, two killed in a hit-and-run accident, and one drowned. Jagger wouldn't play "Sympathy for the Devil" on his next tour.

I Like the Look

Sha-Na-Na almost didn't make it. Famous lead singer Bowser couldn't convince an agent to book the group even though he had a doctorate from the Juilliard School of Music. He came back with his hair greased, wearing old sneakers and dressed in full 1950s regalia. A star was born.

Put the Cat Out

A gold-record singer of the 1970s has given everything up for religion. Cat Stevens, famed for mellow classics like *Teaser and the Firecat*, has become a Muslim and changed his name to Yosef Islam. He said he couldn't cope with his success and is now selling off all his guitars and gold records.

Don't I Know That Bag?

A popular actor-singer had such a tough time breaking into the music business that he helped chance along. David Soul spent two months looking for work in New York before he decided to put a bag over his head, take his picture and send out tapes of his singing. Agents were curious and Soul ended up making 26 appearances on *The Merv Griffin Show* (without the bag) and later filmed the television series *Starsky and Hutch*.

Monkee See, Monkee Do

A popular pop recording star was turned down for a television series because he had bad teeth.

But that initial lousy luck was a blessing in disguise for Stephen Stills of Crosby, Stills, Nash and Young. Stills was turned down for *The Monkees* TV series because his teeth did not photograph well. Ironically, he recommended his friend Peter Tork for the part and Tork got it. But five years later the Monkees were gone, Tork was broke and Stills, now a superstar, ended up buying Tork's Hollywood mansion.

Music Maestro Please

A number of the world's greatest classical composers and artists had lives and deaths ruled by chance. Many of the tales have emerged hundreds of years after their deaths. Some of them:

* Mozart died while composing a mass for the dead. His family was too poor to pay musicians, so there was no music at his funeral.
* Beethoven wrote most of his immortal music while he was stone deaf. He insisted on conducting it at premieres, so the orchestra often got more than a page behind. He had to be turned around to take a bow at the end because he could not hear the applause.
* Brahms wrote his famous Lullaby to comfort a child whose mother had been burned to death.
* Handel was working on the final bars of his oratorio *Jephtha* in 1752 and had just penned the lines "How dark, O Lord, are thy decrees?" when he was struck blind.
* Chopin died for love. He was persuaded to go away with his lover George Sand to her island retreat and the damp air gave him tuberculosis.
* Armand LeBoeuf, a French tenor, sang the role of Romeo with such force that he broke his neck.

I'll Vacuum Tomorrow

Carole King was fascinated when she heard her baby-sitter sing around the house. She insisted Little Eva record the tune and we got "Loco-Motion," a gold record.

And Now, the No. 1 Hit

The most popular song of troops on both sides during World War II almost never made it to the air. "Lili Marleen" is still sung by veterans around the world, but it wasn't supposed to be a hit:

* German cabaret singer Lale Andersen insisted on recording the song in 1939 over everyone's objections. She wanted it to be a backup to the other side of the record, "Three Red Roses," which was being promoted as a hit.

* A drunken disk jockey on Radio Belgrade put on the wrong side of the record one night and started a wave of requests.

* The Germans quickly saw the propaganda aspects of the song and asked Andersen to sing some new anti-American lyrics. She refused and the Nazis interned her until after the war.

The Janis Legend

She may have tempted fate too often. Janis Joplin had all the breaks before her death in 1970. She had just finished her best album, *Pearl*, had a hit single, "Bobby McGee," and had found the love of her life. But she also loved life in the fast lane and died of a drug overdose, apparently self-inflicted.

Rave Reviews

Your chances of being a hit with the critics are not so good in the world of classical music. For example:

* Fritz Kreisler, an eminent violinist, passed off his own tunes as the works of the masters and critics lapped them up. Emboldened, he finally took credit for one of his pieces and the critics accused him of arrogance and lack of any real talent.

* Lashed by poor reviews, Swiss pianist Jean-Jacques Hauser booked himself into a concert, pasted on a moustache and called himself Anton Tartarov. He was given a standing ovation and then tore off his moustache, revealing he had also composed all the music himself. The audience applauded even more loudly.

The Curse of the Opera

The strangest quirk of fate recorded in opera occurred in Paris in 1849.

* Tenor Eugene Massol sang "O God, crush him" during a performance of *Charles VI*. He pointed to the arch of the stage ceiling and a stagehand immediately fell to his death.

* Next night, he sang the phrase to an empty seat. But a patron came in, sat down and fell dead on the spot.

* The third night, Massol indicated the orchestra pit and a musician fell dead.

* The opera was retired for a few years but was later revived to honor Napoleon III. On the way to the theater the emperor and his escorts were bombed by revolutionaries and 156 were injured.

SIR W.S. GILBERT (1836-1911)
OF THE FAMED TEAM OF GILBERT AND SULLIVAN, LEFT A LUNCHEON AT THE DRAMATISTS' CLUB, IN LONDON, ON MAY 26, 1911 - *POINTING OUT TO HIS COMPANIONS THAT HE HAD BEEN THE 13th MAN AT THE TABLE* - 3 DAYS LATER HE WAS FOUND DROWNED IN HIS SWIMMING POOL

The Day the Music Died

Where would they be today if they had lived? Strange things happen in the music business and one of the most amazing is the incredible number of artists who died at the height of their careers. Luck ran out for:

* Buddy Holly, Richie Valens and the Big Bopper (J. P. Richardson). They all died on February 3, 1959, when their plane crashed en route to a concert.

* Otis Redding died in another plane crash in 1967, yet his greatest hit, "Dock of the Bay," came out the next year.

* Sam Cooke was shot to death in his hotel room in 1964.

* Jim Croce died in a plane crash in 1973. Many of his songs, such as "Bad, Bad Leroy Brown" and "Time in a Bottle," are still popular.

* Lynyrd Skynyrd Band was virtually wiped out in another plane crash in 1977, just after releasing an album called *Street Survivors*.

* Jimi Hendrix died in 1970 of an overdose of barbiturates just when he was beginning to experiment with new and fascinating forms of music, as unreleased tapes showed.

* Brian Jones, an original member of the Rolling Stones, was found floating dead in his pool in 1969. He had tried to drown himself twice before but had been rescued by fellow Stone Mick Jagger.

* Jim Morrison, lead singer of the Doors, died of apparent heart failure in Paris in 1971. His wife identified the body, buried Morrison and then killed herself three years later.

* Beatle John Lennon was shot and killed outside his New York apartment by a man who had asked for—and received—an autograph only hours before.

The Writers' Corner

William Shakespeare (1564–1616).

Dante Alighieri (1265-1321).

Fyodor Dostoevsky (1821–1771).

Nowhere in the fields of entertainment are the odds greater for *not* making it than in writing. If you publish a book in North America, you have only one chance in 25,000 that it will sell 100,000 copies.

The odds of getting a book published are somewhat better. The American publisher Doubleday, for example, receives more than 10,000 unsolicited manuscripts a year, out of which three or four get printed.

But writers persevere, take the big chance and finally get their rewards:

* Daniel Defoe took *Robinson Crusoe* to 20 publishers before he finally got it printed. It has been a best-seller for over 250 years and has been translated into 10 languages.

* Milton's epic *Paradise Lost* sold 40 copies on its first printing.

* Dante, author of *The Divine Comedy* and considered the greatest poet of the Middle Ages, wrote songs for a street singer to make ends meet.

* William Shakespeare never had a play published in its entirety during his lifetime. When the plays emerged after his death, they were at first derided as a cheap publicity stunt.

* Feodor Dostoevsky wrote *Crime and Punishment* and most of his other works to pay off gambling debts. He went bankrupt several times and finally fled to Switzerland to escape his creditors.

NOAH WEBSTER (1758-1843)
THE LEXICOGRAPHER WAS INSTRUMENTAL IN SECURING ENACTMENT OF THE COPYRIGHT LAW— YET HIS OWN NAME WENT INTO THE PUBLIC DOMAIN AND COULD BE USED ON REFERENCE WORKS HE DID NOT EDIT

What's in a Name?

A chance discovery of a family journal gave us one of America's best-known writers. But it nearly didn't happen for Zane Grey. A dentist from Ohio, Grey turned to writing to relieve his boredom. Publishers rejected his first book, a collection of diaries from his ancestors called *Betty Zane*. He then turned in *Riders of the Purple Sage* and publishers jumped. More than 50 of his westerns are still in print.

Waiting for the Big Break

Some writers and poets helped chance along, but most sat patiently waiting for success. For many it didn't come until after death, but there were exceptions.

* Robbie Burns turned to writing poetry to raise the money to travel to Jamaica. He opened his own press, ran off 600 copies and made £20. When his next edition of poems made £500, Burns proclaimed himself contemptuous of money.
* Walt Whitman also published his own poems and even set his own type. He guaranteed favorable reviews by writing them himself—in three different newspapers.
* John Bartlett found many of his memorable entries for his *Quotations* by working in a bookstore which he later bought.
* Edgar Allan Poe sold his first book for 12¢. It recently was resold at auction for $11,000.
* Ezra Pound lived on potatoes while waiting for fame. He paid the printer himself for his first book which sold for 6¢ a copy.

The Play's the Thing

A king's insistence that he could write a play finally killed him. Frederick of Germany, unhappy that his sole fame depended on his sword, wrote a play. It was rehearsed twice before the premiere in 1324 when a sword slipped from an actor's grasp and killed the king....

Love Conquers All

The personal life of Mark Twain, who is revered the world over for his writing, was ruled by chance. Twain married Olivia Langdon in 1870, three years after he fell in love with her face on an ivory miniature.

The World's Most Unlikely Writer?

Chances are he'd win hands down. Consider:
* He was expelled from or dropped out of several schools and his writing bore marks of illiteracy.
* He failed to get a $25 raise in 1911 so he turned to writing because he liked to tell himself stories before dropping off to sleep.
* After the only story he ever researched failed miserably, he chose exotic locations so nobody could check on him. He picked the moon or Mars or the jungles of Africa.
* He made up the names of his characters by thinking up syllables and stringing them together.

He's Edgar Rice Burroughs, one of the world's most famous writers and creator of Tarzan.

Death Flight

A set of lost luggage saved a famous writer's life. Novelist Jerzy Kosinski was flying to Los Angeles from Paris when he lost his luggage in New York. He had to get off the plane to retrieve it and missed a party with actress Sharon Tate and her friends on the Coast.

That was the same night that Charles Manson and his followers paid their murderous visit to the Tate house.

Jungle Luck

A nurse who cared for a poor writer's child was paid with a manuscript. Rudyard Kipling told the nurse his manuscript might be worth money some day. Years later, she decided to take him at his word and sold *The Jungle Book* for money that allowed her to live in comfort for the rest of her life.

Hot Stuff

One of the world's classic novels nearly ended up in smoke. The typist working on *Lady Chatterley's Lover* for writer D. H. Lawrence became so upset at some of the steamy passages that she threw five pages into the fire. Lawrence rescued them, but his troubles didn't end there. No publisher would have anything to do with the book. He finally had to send it to a printer in Italy who didn't speak English to get it published.

His Own Obituary

America's greatest humorist left a chilling legacy. Will Rogers, killed in a plane crash in 1935, had partially completed his last newspaper column before he began the fatal flight. The last word Rogers had typed was "death."

Carry On

Joyce Cary became a famous writer because he beat the odds.

Born in 1888, Cary studied painting in Edinburgh and Paris and then took his degree at Oxford. He served with the British Red Cross in the Balkan War and then fought in the Cameroons during World War I.

He was wounded and served in a long series of remote colonial outposts in West Africa, finally returning home in 1920 to write novels.

But not until he turned 44 was his first book published—*Aissa Saved* in 1932. His first critical success came with *Mister Johnson* seven long years later and financial success with *The Horse's Mouth* in 1944.

In 1955, however, Cary learned he was dying of progressive paralysis. He still refused to quit, however, and completed a book of critical lectures before he died in March 1957.

If at First...

The best-selling writer of westerns in the world has seen both sides of chance. Louis L'Amour was destitute at age 12 in North Dakota when his father was forced out of business. He left home at 15 and once earned $500 by fighting a 230-pound Irishman and knocking him out in the sixth round. That tough early life prepared him for the 26 consecutive rejections of his first novel. He finally sold it and has gone on to sell more than 100 million books, 80 titles and 32 movies.

A Deceiving Ghost

Several publishing houses took a chance on Clifford Irving and now want to forget it. Irving presold a ghostwritten autobiography of billionaire Howard Hughes and got more than $1 million in advances. The work was exposed as total fiction and Irving went to jail for 17 months. He had to pay back all the money he'd received and went bankrupt in 1976 with $700,000 in legal fees.

HENRY
DAVID
THOREAU
(1817-1862)
WROTE
20 VOLUMES, YET HE NEVER EARNED
A LIVING BY HIS PEN ~ HIS FIRST
BOOK SOLD ONLY 219 COPIES AND
HIS FAMED "WALDEN" DID NOT SELL
OUT ITS FIRST EDITION OF 2,000
COPIES FOR 5 YEARS

The Bear Facts

A sentimental gift on a lonely Christmas Eve helped create one of the world's most delightful children's characters. London writer Michael Bond was on his way home in 1956 when he chanced on a lonely little toy bear left unsold the day before Christmas. He bought it for his wife, named it after his subway stop and later wrote about it. It was called Paddington Bear and the rest is history, especially if you're 5 years old.

Try and Try Again

One of America's most beloved writers was rejected 20 times by the magazine that eventually bought most of his work.

James Thurber started writing sketches for the *New Yorker* in 1926, but they kept turning him down before finally accepting a short piece on a man caught in a revolving door. Thurber never looked back.

He published more than 20 books of collected prose and delightful pictures he drew himself.

Thanks, but No Thanks

If writers had relied on critical opinion instead of taking a chance, we would not have had:

* *History of Mathematics*. This reference work has sold more than 100,000 copies for James Newman despite the fact that it was rejected out of hand by Harper Bros.

* *Ordinary People* was turned down cold by Random House. Judith Guest then went to Viking and the book became a Book of the Month Club Selection, a paperback blockbuster for a $1.5 million advance, and an Academy Award-winning movie.

* *Jonathan Livingston Seagull, Love Story* and *All Things Bright and Beautiful* were rejected at least 12 times each.

* John Creasey, who has published 543 mysteries under 25 pen names, was rejected by 743 editors before he caught on.

On the other hand, medical opinion has contributed to the creation of some classic works:

* Anthony Burgess wrote *Clockwork Orange* and eight other novels after he was told he had a year to live.

* S. S. Van Dine, creator of detective Philo Vance, was sent to a sanatorium where all he could find to read was detective fiction. He decided he could do better and did.

Jail Stories

A prison term gave us the stories of Marco Polo.

When he returned from his historic 22-year adventure in the Far East in the 1290s, Polo became commander of a war vessel. He was captured in a battle off Curzoid Island and jailed in Genoa for a year. While in jail he met a writer named Rustichello, who insisted Polo write down his Far-Eastern stories.

Food for Thought

If Bram Stoker hadn't eaten crab, the chances are we'd never have had one of the world's most famous horror stories. Stoker got the idea for *Dracula* after a nightmare brought on by eating crabs at dinner. Just think what we'd have missed if he hadn't had that dream.

The Last Train

The death of his best-known heroine pre-figured the death of this famed Russian writer. Leo Tolstoi, who some have called the greatest of all novelists, died in 1910 in a railway station like the ill-fated heroine in his novel *Anna Karenina*.

High Pay

Science fiction master Ray Bradbury, author of *The Martian Chronicles*, sold his first story at age 20 to *Script* magazine in Los Angeles.

He received no money—just a year's subscription to the magazine.

You Are What You Eat

One of America's most exciting writers and movie directors has peculiar eating habits.

Michael Crichton, a Harvard-educated doctor, the author of 38 books including *Andromeda Strain* and *Terminal Man*, director of three money-making movies, has a thing about lunches.

Crichton ate nothing but tuna fish during *The Great Train Robbery*, salad and barbe-cued beef while writing *Congo*, and open-faced turkey sandwiches for the screenplay of *Looker*.

He admits he would like to wear the same clothes every day but is concerned they'd smell.

A Writing Machine

Author Georges Simenon defied chance.

Born in Belgium in 1903, he published his first novel at 17 by writing it in 10 days. He used 16 different pen names and began writing hundreds of commercial novels—one of them in exactly 25 hours.

He was most famous for his series of books about Inspector Maigret, but he also delighted in writing a tense psychological novel of less than 200 pages—known throughout Europe as a simenon.

In all, he has published more than 150 novels under his own name and 350 under various pseudonyms.

CHARLES DICKENS (1812-1870) CONCEIVED "PICKWICK PAPERS" AS A PICTURE BOOK WITH ABBREVIATED TEXT, BUT THE ARTIST COMMITTED SUICIDE AND *DICKENS WAS FORCED TO ENLARGE HIS TEXT—WHICH MADE HIM FAMOUS AT THE AGE OF 24*

Nothing's Easy

William Faulkner failed to graduate from high school because he didn't have enough credits.

He bummed around the United States and Canada, enlisting in the Royal Canadian Air Force, trying to get into university and later working as a postmaster until he was fired for reading on the job.

He then tried writing and had five books finished by 1930 but failed to earn enough money to support a family. But he kept going and became popular in the mid 1930s. He eventually received the Nobel Prize for literature in 1949.

Try and Try Again

Nobody had worse luck with publishers than James Joyce. He threw his first auto-biographical novel, *Stephen Hero*, into the fire when they rejected it, but went on to write *Ulysses*. It was considered a dirty book, however, and was banned in America until 1933, more than 10 years after its publication.

JULES VERNE
REGARDED AS THE FATHER OF SCIENCE FICTION, WROTE A SCIENTIFIC TREATISE ON BALLOONS BUT IT WAS SPURNED BY PUBLISHERS UNTIL HE OFFERED IT AS A FICTIONAL ADVENTURE TITLED: *"FIVE WEEKS IN A BALLOON"*

Bonjour Francoise

Francoise Sagan began writing in 1952 after she failed her examinations at the Sorbonne.

Her first book was called *Bonjour Tristesse* and it sold more than 700,000 copies in France alone. It eventually was translated into 14 languages and her next two books sold more than two million copies in America.

Nobody's Perfect

The world's first "perfect book" was published in 1744 only after six proofreaders read every page 500 times. A reward of $250 was posted for each error discovered, but when the edition of Horace was finally printed, six typographical errors were found—one on the first line of the first page!

If You're Out There, Flush

Chances are six in 10 that this book will be read in a bathroom.

I Know What I Like

What is art? In many cases, a chance review, word of mouth or an untimely death elevates art into high art. Sadly, many of the great masters were not recognized until after they died, but on the other side of the coin, many modern painters have lived rich and famous lives.

Here are some classic tales of chance:

* Auguste Rodin, who did *The Thinker* and *The Lovers*, was rejected three times by the Ecole des Beaux-Arts for lack of talent.

* Van Eyck's priceless portrait of his wife was found in a fish market in Bruges, Belgium, in 1808. It was being used as a tray to display fish.

* Vincent van Gogh's portrait of Dr. Rey, painted while he was in a sanatorium and given to the doctor as a token of thanks, was found in a crack in a chicken coop. Van Gogh sold only two paintings in his lifetime, for a total of $84.

* A chance win in a lottery gave a famous French painter the freedom to pursue his art.

Claude Monet won 100,000 francs in 1891 and spent the rest of his days wandering about and painting the French countryside.

Maybe She's Happy

Art critics have been trying to figure out the enigmatic Mona Lisa smile ever since Leonardo da Vinci painted her. An expert from Copenhagen diagnosed it as facial palsy, and a British doctor said she was pregnant, adding that she wasn't wearing a ring. Regardless, the painting is among the most famous in the world and when it appeared in the Metropolitan Museum in New York, each visitor was allowed only a 10-second glimpse.

And They Took Money for It

Chance dealt the masters a cruel blow, but it also blows the other way today. The following items have all been considered fine art:
* A pile of bricks on display in the Tate Gallery in London.
* A work called *Room Temperature*, with two dead flies in a bucket of water.
* A sculpture called *Stone Field*, made up of 36 boulders.
* Photographs of 650 San Diego garages.
* A covered piece of metal, with a hole so the viewer can watch it turn green.

Ditchdigger

Only chance let Paul Gauguin live.

For a short time he was a common laborer on the Panama Canal and worked 13-hour days digging under the tropical sun and rain. More than 25,000 workers died during construction, but Gauguin lived until 1903, leaving behind a legacy of beautiful paintings and woodcuts.

JAMES McNEILL WHISTLER (1834-1903)
AFTER HIS MOTHER'S DEATH, RETRIEVED HIS FAMOUS PAINTING, "WHISTLER'S MOTHER," FROM A PAWNSHOP BY BORROW- ING 50 POUNDS -- *BUT WAS UNABLE TO SELL IT LATER IN NEW YORK FOR $500*

Life after Birth

A near fatal misjudgment marked Pablo Picasso's birth. He was abandoned on a table when a midwife thought he was still-born. But an uncle noticed the baby struggling for breath and fetched a doctor, saving his life.

When he died in 1973, Picasso left behind 1,876 pictures, 1,355 sculptures, 2,880 ceramics, 11,000 drawings and sketches and 27,000 etchings and lithographs. His estate was worth $250 million.

What's 200 Years Anyway?

Junk? Hardly. This is some of the work of the famous artist Jasper Johns.

America's Bicentennial entry in the Venice art exhibition was a stick the size of a pencil, three quarters of an inch thick, unpainted and standing alone. It had no title because the Americans wanted to evade charges of elitism.

Weighty Thoughts

Strapped for money, one of the world's best-known jewelers had to flee city after city because he took a chance with his metals. Benvenuto Cellini, recognized as the Renaissance master in gold and silver design, used base metals like lead and then dipped his pieces in precious metals. He was caught when his masterpieces didn't come up to full weight.

I'll Be Back Tomorrow

One of the greatest artists of all time had trouble finishing anything. Leonardo da Vinci started works throughout Italy, but left much work unfinished when his commissions ran out.

Oops

Le Bateau, a painting by Henri Matisse, was displayed for 47 days in 1961 at the Museum of Modern Art in New York before someone finally noticed it was upside down.

More than 120,000 people had passed by it before the mistake was noted.

It's a Crime

Chance plays a big role in the world of art fraud and fortune. Even the great Michelangelo buried his first sculptures in the backyard so he could age them and pass them off as antiques. Then there are the following:

* In Brunswick, West Germany, an artist painted and signed a fake Monet over a painting of two nudes. He painted two more nudes over the fake Monet, took the whole thing to a dealer to be cleaned and allowed the "masterpiece" to be discovered.

* David Stein copied the works of dozens of artists from Renoir to Gauguin. Marc Chagall turned him in when he saw copies of his own works. But Stein emerged from jail after 16 months to set up his own art-fraud squad and became well known for his own paintings.

* The Cleveland Museum declared a wooden Madonna and Child to be authentic 13th-century Italian sculpture. It was actually done by art restorer Alceo Dossena. The fake was unmasked in 1928 and the museum bought another piece to replace it. It was done by Dossena too.

* A pre-Roman statue bought by New York's Metroploitan Museum in 1918 for $40,000 was actually the work of Alfredo Fioravante, who confessed to the forgery in 1960.

Yours or Mine?

An art forger was so skilled that he fooled the artist he was copying.

Claude Latour excelled in copying Maurice Utrillo, who once confessed that he often wasn't sure which were his and which were Latour's.

"BRICK KNOT" A WORK OF MODERN ART BY WENDY TAYLOR EXHIBITED AT A GALLERY IN LONDON, ENGLAND

PART TWO:

The Wheel of Fortune

Bets, Contests and the Sporting Life

*L*ady Luck has brought un-dreamed-of wealth to gamblers throughout history; to others she has brought misery, despair and even violent and bloody death!

Yet gambling is in our blood. We all dream of that million-dollar win on the lottery, of drawing the un-beatable hand in poker and bridge.

But what are the mathematical chances of winning the big one? Are there born winners as well as born losers?

You'll meet them all in this chapter —fortunes won and lost on the turn of a card, the kingdom that changed hands on the throw of dice, the most amazing bets the world has ever seen, the lotteries that helped build a nation.

Read about:

* The blackjack sharks who haunted the casinos of the world.
* The most horrific card deck of all time.
* The stately home in England that changed hands—because of snails.

* The man who walked 478 miles— literally for peanuts.
* The brutal commander who played chess with his prisoners.
* The most unbelievable horse races in history and some of the fantastic wagers that have been made: the horse that won by a tail, the dray horse that became a race horse, and the jockey who was arrested for speeding.
* The high rollers of the world's casinos: the great actress Sarah Bern-hardt and The Man Who Broke the Bank at Monte Carlo.
* The King of Clubs in Britain, the flamboyant John Aspinall.
* The world's greatest lotteries and the small town in Spain that won a fortune.
* The lotteries that helped build the New World.
* The math error that netted a fortune.

Take your seats then for gambling and chance. It's the biggest game in town—and Luck is your partner.

Bridge Player's Dream

In bridge, a player must get all 13 cards of one suit for a "perfect hand." The odds against it? 158,753,389,900 to 1. Yet a perfect hand of diamonds was dealt to J. E. Wacksmuth in Pasadena, California, in 1933.

The odds against a perfect deal are 2,235,197,406,895,366,368,301,560,000 to 1. Yet on Christmas Day, 1965, all four players got perfect hands. Mrs. Jessie Cun-nington of Louisville, Kentucky, dealt Mrs. Eugenia Shontz 13 clubs, Mrs. Bruce Long 13 diamonds, Mrs. Richard Black 13 hearts and herself 13 spades.

And in a game in Trenton, Illinois, in 1962, Mrs. Roscoe Nenz was dealt 13 clubs, Mrs. Roland Tschudy 13 diamonds, Mrs. Floyd Tschannen 13 hearts and Mrs. Richard Hammel 13 spades. Incredible, but true!

A Fair Deal—The World of Cards

Take a chance, pick a card. Whether you play in million-dollar casinos or your own home, whether it's high-stakes blackjack, poker or stud, rubber bridge, whist or pinochle, sooner or later you'll meet chance face to face.

Fortunes have been made or lost on the turn of a card. Gamblers devote their lives to trying to master a certain game or a certain bet. Chance sits on every hand and luck is your partner.

Cards have fascinated and infuriated man for centuries, and it all began with money. The Chinese are generally credited with inventing the first card game back in the seventh century, using different combinations of money which they shuffled into a variety of hands.

Games arrived in Europe in the 13th century and were first manufactured in England in 1486, during the reign of Henry VII. His wife, Elizabeth of York, has the best-known face in history. Her likeness appears eight times on all decks of playing cards manufactured since her day!

The first playing cards in the New World were made of leaves. The sailors on Columbus's ships were inveterate gamblers, but threw their cards overboard in superstitious terror during a storm. When they landed in America, they immediately set about making new cards for themselves, out of the leaves of the copas tree!

Now, 1,300 years after the Chinese, we're still playing with cards. But, as we shall quickly see, some people do it a lot better than others.

Poker: The Big Draw

Poker. The name conjures up the Wild West, blazing six-guns and death. When it is played with five cards, there are 2,598,960 possible hands in a 52-card deck. It all depends on the luck of the draw. Some hands have brought fame and fortune, others disgrace and murder most foul.

* Dead man's hand. Wild Bill Hickok was murdered while playing poker. He was holding three aces and a pair of black eights when another gunfighter walked behind him and shot him through the back of the head. Hickok's poker hand has been known ever since as the "dead man's hand."

* Flushed with success. Edwin P. Jones of Philadelphia held two royal flushes in one evening in 1923. Six years later he held another—in the same house with the same players. Odds against a royal flush happening once are 122,115,869 to 1.

* Four aces lost! A straight flush was dealt to F. Gladden Searle of Connecticut in an incredible seven-card-stud poker game in 1956. The other players held four aces, four kings and four queens.

Dead Man's Shoes

The strangest inheritance in history. Robert Fallon of Northumberland, England, was accused of cheating in a poker game in San Francisco's Bella Union Saloon in 1858—and was shot dead. He'd won $600.

The other players believed the money to be unlucky and called in the first passerby to sit in for the dead man, confident he would soon lose the $600.

The young stranger had increased the sum to $2,200 by the time the police arrived. When the police demanded the original $600 for the dead man's next of kin the youth proved he was the unlucky gambler's son—who had not seen his father for seven years!

The Biggest Legal Game of All

Forty-two players sat down in Las Vegas in May 1979 to play $10,000 buy-in poker with no limit a hand. But as chance would have it, the final two were veteran Crandall Addington, who had played for 10 years in the world series of poker, and brash young Bobby Baldwin, out of Oklahoma State University.

With a $210,000 first prize riding, Addington and Baldwin went head to head for four days until Baldwin asked for a break and went up to his room to change his clothes. He returned in his lucky apple-green shirt and the game switched to hold 'em, a deceptively simple version of seven-card stud with five community cards.

Baldwin, his lucky shirt radiating, started to push and built the pot up to $92,000 when the fourth card turned up was an ace. Crandall hesitated and finally folded—and then Baldwin showed him his hole cards—a nine and a ten. It was a $100,000 bluff and Crandall never recovered.

Luck or skill? Bobby Baldwin would like to believe it was a bit of both.

Double Down

So, you like to play a little blackjack. Well, the odds are against you. The odds against pulling a natural 21 in two cards are 20.1 to 1, but the odds against getting 20 are only 8.4 to 1.

Chances of getting a two-card combination of 20 or better are 5.5 to 1 against you, while 16 or better is only 1.4 to 1.

X Marks the Spot

Cross-eyed crooks made a fortune in cheating highly paid construction workers in Canada. The Royal Canadian Mounted Police tell of one scam in 1969 involving hair dye applied to the backs of cards and then quickly wiped off. But the residue of the dye can be seen by throwing one's eyes out of focus, a trick that can be learned in 30 minutes. The RCMP estimated roaming bands of cardsharps took an average $12,000 a night in remote camps.

Turns of Fate

Fortunes have been won and lost on the luck of the draw. Just consider some of these examples:

* The duchess of Mazarin (1646–1699) inherited more than £5 million—then lost every cent of it at cards!
* The duke of Guise (1519–1563) wagered his entire fortune—more than £8 million—on the turn of a single card. And he gave fantastic odds of 400 to 1! He won, earning a mere £20,000—which he gave to the messenger who brought him his winnings.
* The castle of Kerousein in France changed hands in a single game of cards in 1880.
* Henry Bosville of England lost £125,000 at cards in 1750 and won it all back the next year. He never gambled again.
* The celebrated British statesman Charles James Fox (1749–1806) lost more than £500,000 at cards before he was 21.
* William Northmore (1690–1735) of Okehampton, England, lost £400,000 on a single card—and was elected M.P. in 1714 by his neighbors who were sorry for him!
* The National Library of France was won on the turn of a card. The famous statesman Cardinal Mazarin was playing piquet in 1644 with Jacques Tuboeuf, who stood to win 200,000 francs. Mazarin won the building

The dying Mazarin, Premier of France, played a last game of cards on his deathbed, the loser to pay the funeral expenses. Mazarin died as soon as the game was finished. Believe It or Not!

Charles James Fox.

which was bought by the king in 1721 and became the home of the library.
* The Hotel Sully in Paris, France, was lost before it was even built! The loser in the card game in 1624 had to construct it for the winner.

Double Down and Deal Again

More and more people are trying to take the chance out of blackjack. Casinos in Las Vegas call them counters—men and women who remember high cards no matter how many decks are used—and try to ban them, thus retaining the house edge. But some players succeeded in eliminating chance and made a fortune before the casinos got wise.

* Ken Uston headed a team that made $145,000 in nine days at Las Vegas. A Czech team made double that in Atlantic City and a New York computer programmer named Arthur Peyser picked up $70,000. The casinos then banned counting.

* Uston liked to talk about his big scores and said his biggest payday was $25,700 at the Fremont in Las Vegas. Even on slow days he averaged $400 an hour profit or $4,000 for a full shift. He reckons his edge against the house is 1 percent, so he sometimes has to play 11 or more hours to win.

* So devious was Uston that he used to send in a trained team when the casinos banned him. He estimates his team placed $600 million in bets and won it back, along with $3 million in profits.

* Uston is suing two casinos in Nevada for infringing on his right to gamble, even though he admits he developed a computer he wore in his jockstrap and a foot-telegraph signal in his heel.

Stacked against You

You may be a darling of fortune but it'll do you no good at many card games. Card manufacturers say one of every 10 games of cards played for high stakes is played with a marked deck!

Five Beats Four

In 1981 a poker player was shot dead when a fifth ace fell from under a card table during a game with a $27 pot. Host James Lewis was charged with murdering Arthur Ellison, who was shot four times in the back and once in the head in Winslow, New Jersey.

A Chance Snack?

How's this for dedication?

The earl of Sandwich resented leaving the card table each night for dinner so he invented the sandwich.

Ugh

The most gruesome card deck of all time was discovered by an Indian trader in Arizona.

Capt. E. M. Kingsbury decided to play in a chance game at the San Carlos Indian reserve. He noticed the cards had painted faces and were made of some thin material, more durable than rawhide.

After playing for an hour or so, Captain Kingsbury discovered the cards were made of tanned human skin—the flesh of captured white men.

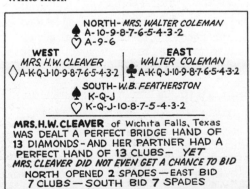

NORTH– *MRS. WALTER COLEMAN*
♠ A-10-9-8-7-6-5-4-3-2
♡ A-9-6

WEST
MRS. H.W. CLEAVER
◇ A-K-Q-J-10-9-8-7-6-5-4-3-2

EAST
WALTER COLEMAN
♣ A-K-Q-J-10-9-8-7-6-5-4-3-2

SOUTH– *W. B. FEATHERSTON*
♠ K-Q-J
♡ K-Q-J-10-8-7-5-4-3-2

MRS. H.W. CLEAVER of Wichita Falls, Texas WAS DEALT A PERFECT BRIDGE HAND OF 13 DIAMONDS–AND HER PARTNER HAD A PERFECT HAND OF 13 CLUBS– *YET MRS. CLEAVER DID NOT EVEN GET A CHANCE TO BID* NORTH OPENED 2 SPADES—EAST BID 7 CLUBS—SOUTH BID 7 SPADES

Rolls of the Dice

Kingdoms have changed hands on the throw of a die. Fortunes have been won and lost. For some people it's a matter of life—or death.

Believe it or not, we've been dicing with fate since the earliest times. Dice made from the ankle bones of sheep are among the earliest human artifacts, and the game of craps is the modern version of a game played by Roman soldiers.

The die is cast. Here are some examples of the fickle roll of the dice:

* A kingdom won on a lucky streak: in the early 11th century King Olav of Norway and King Olof of Sweden rolled dice to decide who would own the island of Hisingen. Both threw double sixes six times in a row! Finally the Norwegian king hurled down the dice with such force that one split—and Hisingen has been Swedish ever since!

* Dicing with death: Gen. Adam Herberstorf of Austria devised a fiendish punishment for peasants who rebelled in Poland. He divided the 16 rebel leaders into two groups and ordered them to dice against each other. It was winner take all: those who won went free, those who lost were hanged.

* Death was not the end of the game for Flemish doctor Pquier Joostens (1535–1590). He was such a fanatic about dicing that he ordered his own skin to be used as the cover of a dice board and his bones to be made into dice!

Loaded against Them

They cheated at Pompeii—and lost! Loaded dice have been dug up in the Roman city of Pompeii, destroyed in 79 A.D. when Vesuvius erupted.

Unlucky Break

The fire that destroyed Chicago started because of a game of dice. It wasn't Mrs. Patrick O'Leary's cow that kicked over that lantern, as the popular story goes. A man shooting craps in her barn knocked it over accidentally.

Wanna Bet?

Horace Walpole, 4th Earl of Orford.

People have made some fantastic and bizarre bets on anything that moves—from horses to raindrops and snails.

They've even bet on things that don't move. Witness this story by the famous English writer Horace Walpole (1717–1797) about an incident at White's, a London club for gentlemen:

* A man was carried into White's in a state of collapse and bets were made on whether he would live or die. When someone suggested calling a doctor, it was rejected since it would compromise the bet.

"He was therefore left to himself," wrote Walpole, "and presently expired—to the great satisfaction of everyone who had bet for that event."

* Lord Arlington (1618–1685) wagered £3,000—today equivalent to £150,000—on which of two drops of water would reach the bottom of a windowpane first.

* Godolphin House in Helston, England, and its entire estate were lost to the St. Aubyn family in a bet on a snail race. Because he won on a technicality, the winner allowed the Godolphins to stay in the house—for an annual rent of just 30 pence (66¢).

* Ann, Lady Cobham, wife of Sir Richard Temple, lost £50,000 in one race in 1760—betting on worms!

Getting Their Deserts

Dice made from the hooves of animals are always carried by the Masarwas of the Kalahari Desert in South-West Africa—to be thrown for a divine decision on any subject.

Love's Labor's Lost

Greek millionaire Giorgios Averoff offered his daughter's hand plus a dowry of $193,000 to any Greek who won the marathon when the Olympic Games were revived in 1896. Spiridion Loves, a Greek, won the marathon—but he was already married!

General Lost—and Won

Sir John Cope, dismissed as English commander-in-chief after he lost the Battle of Prestonpans to the Jacobite Highlanders in 1745, bet every penny of his savings—£25,000—that his successor would do no better. The following year Gen. Henry Hawley took on the Scots again—and was whipped at the Battle of Falkirk.

Walking for Peanuts

Edward Payson Weston walked from Boston, Massachusetts, to Washington, D.C.—a distance of 478 miles—in 10 days in 1861. He won the wager—a bag of peanuts.

A Piece of Cake

Squire John Rowers bet $10 he could eat a more expensive breakfast than a friend. He then put a $100 bill between two pieces of cake—and ate it.

A Fast Operation

Gamblers will do anything to keep on betting. One man woke up from an appendectomy and sneaked out of the hospital to place a bet—still bleeding.

They're Off!

A bus bound for Mohawk Raceway near Toronto overturned in March 1978. Even injured passengers cheerfully climbed out and thumbed rides to the track. All had disappeared by the time the police arrived.

Take the Stage

One of the oddest bets in history involved a stage and a steamboat. In 1859 John Butlerfield and his Overland stage left St. Louis for San Francisco at the same moment the Great Eastern steamship departed New York for San Francisco via South America.

The bet was $100,000 and Butlerfield, using relays of mules, covered the 2,795 miles in 20 days. He beat the steamship by 36 hours.

I'll Bet on It

More and more people around the world are taking a chance on gambling. It can be old Dad tottering off to the betting shop in Britain or someone staking a few thousand lira in the government casinos of Italy. A brief rundown:

* Britain has 125 licensed casinos and £750 million is staked yearly. There are more than 1,000 licensed gaming houses, and licensed bingo halls number 1,687. From these establishments, plus slots, football pools, and horse and dog racing, the government collected more than £300 million in 1972 alone.

* The tiny principality of Monaco (population 24,000) has been entirely supported by gambling since 1861. The famous Monte Carlo casino makes an annual grant to the government and it's so big that there's no need for income tax.

* Lotteries have been in existence in West Germany for 100 years and the government averages more than $250 million from that action alone.

* All gambling in Italy is administered by the government or state-run bodies, and the most popular bet is on soccer. Revenues run around $500 million and illegal betting is virtually nonexistent.

* But in the United States, illegal bookmaking on National Football League games is estimated to run around $767 million yearly.

* Manila has closed its gambling casinos, but cockfighting is still legal. It's privately run, with an enormous amount of money bet weekly.

The Laws of Odds

In gambling, the odds are always against you. But it helps to know what the laws of chance are:

∗ All chances lie between zero (no chance) and 1 (a certainty). An even bet is 0.5. Any event either will or will not happen.

∗ If only one event among several can happen, and if all are equally probable, the chances of a particular event happening are one in the number of possibilities. For example, the chances of rolling a four on a die are one in six.

∗ The probability of an event happening more than once is found by multiplying the individual odds together. For example, the chances of rolling a four twice in a row are one in 36, three times in a row, one in 216.

Still, chance manages to evade the laws of probability. For example, in any toss of a coin, the likelihood of heads coming up is even. But on several throws, heads may come up several times in a row. This is merely a small "lucky streak." The more often you throw, the likelier it is that you will throw heads about half the time. But there are no guarantees.

In gambling casinos, for example, numbers come up either black or red (discounting the house number 0). The odds of black occurring and of red occurring are the same. But in Monte Carlo on August 8, 1913, black came up 26 times in a row. If a player had bet $4 on black the first time it came up and let his winnings ride, he'd have won more than $250 million.

What did gamblers do on this occasion? They bet more and more money on red, confident it would turn up soon. After black had come up for the 20th time, they thought there was no longer a chance in a million that the run would continue. But it did—and the casino won millions.

Shortening the Odds

Bridge expert Charles Goren has a word of advice about betting: don't!

But if you have to bet, consider this: of all the casino games, only blackjack favors the bettor, and then by only 1 or 2 percent! As for the rest, they operate on the W. C. Fields principle: "Never give a sucker an even break!" The odds against you at the roulette table, for example, are more than 75 to 1!

In horse racing, an eight-horse race offers odds of precisely 8 to 1—no matter which horse you bet on.

Microchip Man

Keith Taft, a devout, Bible-reading Baptist, uses microchip technology to rake in the chips. He hides his space-age minicomputer in his jockstrap and reckons it gives him a 2.5 percent edge. David the computer can whirl through 100,000 calculations a second!

That Sinking Feeling

Scottish legend has it that Glamis Castle, the birthplace of Princess Margaret, is actually the tomb of inveterate gamblers.

The superstition concerns a nobleman and his four associates, who were playing in the castle when a clergyman arrived and criticized them for evil practices. They swore they would continue playing "until the end of the world."

But Scots swear that immediately afterwards, the room in which they sat sank under the castle. Legend has it that the gamblers still play and on dark nights their voices can be heard within the stony walls....

Horse Racing: Riding the Odds

Horse racing. It's one of the oldest sports in the history of man. It's called the sport of kings—and kingly fortunes have changed hands on racecourses round the world. Few other sports can match it for heroic deeds, bizarre coincidences and quirks of fate!

Not Such Dumb Animals

The earl of Glasgow, who had lost millions betting on his own horses, announced he would shoot the next one that lost. That afternoon, on October 30, 1852, all six of his horses won!

This One Will Run and Run

The Great American Horse Race, from Frankfort, New York, to Sacramento, California, was won in 1976 by Virl Norton of San Jose, who rode the 3,200 miles in 315 hours and 47 minutes—on a mule named Lord Fauntleroy.

Oh Brother!

In a race in Havana, Cuba, in 1941, two horses owned by the same breeder and belonging to the same entry were ridden by two brothers. They finished in a dead heat!

Win, Place, Show

Jockey Eddie Arcaro, riding in six races at Belmont, New York, on June 7, 1957, finished first, second, third, fourth, fifth and sixth!

Dead Heat, Dead Heat, Dead Heat

In a race in Sydney in 1903, there was a triple dead heat. The three horses—High Flyer, Loch Lochie and Bardini—ran the heat over again and the result was another triple dead heat!

Bet on the Bay

One of the most amazing racehorses of all time was Theodore, a bay owned by Edward Robert Petre. He won the St. Leger race at Doncaster, England, in 1822, although he was so lame the betting against him was 5,000 to 1. Jockey John Jackson had wept when ordered to ride Theodore, yet the horse—still lame—went on to win major races at York, Manchester and Edinburgh!

Horse before the Cart

Pasha, an 8-year-old dray horse that pulled a bakery wagon in Willheimshaven, Germany, was entered in a local race in 1912—and won easily!

Speed Trap

A horse race was staged in Seattle, Washington, on July 18, 1881, between Tom Clancy and Robert Abrams. City authorities cleared the site for them—but the winner was arrested after the race for breaking the speed limit of 6 mph.

Good Start

Joe Notter won his first race as a jockey in 1904—at odds of 100 to 1.

Hot Favorite

Black Tommy, an English Derby horse of 1857, was such a prohibitive favorite that his owner offered a bet of £50,000 and could only scare up a coat, a hat and a vest. Black Tommy—and his owner—lost.

There's Many a Slip...

Chance affects all horse races, but this thoroughbred really pushed his luck.

The horse Brampton was entered in a race in Dargaville, New Zealand, on February 2, 1931, and led almost to the finish. Jockey Joe Parson urged him on only 40 feet from the finish line when Brampton suddenly fell.

But he rolled several times and—with Parson hanging on for dear life—slid across the wire to finish first by a tail.

A Sure Loser

Senacas Coin was probably the worst horse ever to run in the Kentucky Derby. And chances are no one will ever duplicate his record.

In 1949 he pulled up a quarter of a mile behind the victorious Ponder. His jockey, Jimmy Duff, had to ask who'd won the race when he finally returned to the barn. Senacas Coin continued his career and started 53 times, losing 52.

Dead Heat

An extraordinary 2 dead-heats, 2 years apart, with the same 2 horses, Harvard Square and Mettlesome!

ANGEL CORDERO Jr.

IS THE ONLY JOCKEY WHO EVER WON MORE THAN $5,000,000 IN PURSES IN A SINGLE SEASON THREE TIMES

Wheeling and Dealing

The world's great casinos are a fool's paradise, right? Wrong. They're the most honest place to lose your money; local games are far more likely to be rigged.

Here's a glimpse into the world of the great casinos at Las Vegas and Monte Carlo, with their exotic games of baccarat, roulette and blackjack.

* The casinos' greatest fear is suicide. Francois Blanc, Monte Carlo's founder, invented a special system to head off potential trouble.

The croupier presses a button and a special "suicide squad" whisks the poor gambler away, while someone else takes his place at the table.

* It's okay for visitors to Monte Carlo but residents of Monaco are forbidden to bet at casinos or even to enter them!

* The average Swiss bets less than $20 a year. Swiss law bans bets of more than $2.

* The only time gambling stops at Las Vegas is for half a minute on the stroke of midnight on New Year's Eve.

The Ape of Spades

Gambling skills learned at Oxford helped produce one of Britain's best-known club owners and animal lovers. So incredible was the life of John Aspinall that he once risked everything on a long shot and put his new-born son into the arms of a gorilla.

Consider:

* Aspinall spent his years at Oxford gambling and learning to write poetry. He became so accomplished at the former that he lived on a scale far grander than any of his fellows. He liked to wear magenta suits and pink shirts and carry a cane.

* He built and lost a fortune quickly by kiting. Realizing it took three days for checks to clear, Aspinall wrote them to finance his games. All went well until he hit a losing streak.

* Undaunted, Aspinall took his expense money from a job in Nigeria and hit a 300-to-1 shot at the track in 1953, setting himself up with £7,000. He quickly opened London's first floating chemin de fer game in an apartment in Mayfair and built another fortune.

* In 1962, with gambling laws relaxing in Britain, Aspinall opened a permanent spot which quickly became famous for a standard of decadence unequaled even by Monte Carlo.

He liked to festoon his staircase with dwarfs, while wild animals roamed around the rooms and acrobats hung from the chandeliers. He made another fortune.

* By 1970, however, the laws had changed again and owners were no longer allowed to play. Aspinall got out and began a wildlife preservation campaign. He stocked his house with tigers, gorillas, rhinos, wolves and leopards and became a sought-after speaker.

* After nine years at £500,000 a year, Aspinall decided he could not support his wildlife endeavors, so he returned to gambling and opened one of the biggest clubs in London, featuring a £10,000 roulette limit.

* He lost £1.5 million immediately, but chance stepped in again and Aspinall won it all back and more. His club stands today.

The Man Who Broke the Bank

Fortune favors the bold. And Charles Wells, the paunchy, bald cockney who achieved worldwide fame as "The Man Who Broke the Bank at Monte Carlo," had all the nerve in the world. But finally even his extraordinary luck ran out.

He'd made a fortune in Britain selling false patents before arriving in Monte Carlo in the late 1880s. Yet he started in Monte with only £200. But he won so much so quickly that his daily winnings were published in newspapers around the world.

Each roulette table carried reserves of only 100,000 francs, and Wells "broke the bank"—sent the cashier to the vaults for more money—a phenomenal three times in one year, 1887.

His second visit was equally successful— he won 98,000 francs with an incredible run on the number 5, after starting with a stake of only 120 francs.

But the third visit was a disaster. Wells lost his backers' money as well as his own, and back in London, Scotland Yard began investigating him. In court Wells asked, "How could a man with my system of winning stoop to fraud?" But when neither he nor his lawyer could describe the system, he was jailed for eight years.

He died flat broke in Paris in 1926.

You Win Some, You Lose Some

* A tiny flaw in the Monte Carlo wheel was spotted by British engineer Charles Jagger. They fixed it—but not before he'd won $100,000.

* Lawyer Jeff Randolph, a bachelor from Delano, California, hit the jackpot in Nevada in July 1981. Not only did he scoop $992,012.15 with a $3 bet at a slot machine—he received seven proposals of marriage!

* Frank Lee Lorraine's name came up on a $2 bill in Mexico City in 1945. It was the same bill he'd written his name, address and phone number on and spent in Philadelphia—20 years earlier!

* Professional gambler Nicholas Zugraphos (1886–1953) left an estate of $14 million. He'd won every cent of it playing baccarat!

* Harry Gordon Selfridge, millionaire owner of the first department stores in Britain, gave the Dolly Sisters—a popular dance act—more than £2 million to spend in Monte Carlo. They lost it all.

* Sarah Bernhardt, the famous actress, was a fanatical gambler and once lost more than $100,000 in one night at Monte Carlo. She took an overdose of sleeping pills and nearly died. She'd squandered her life savings—and she never went back again.

Mr. Harry Gordon Selfridge at St. Moritz, 1924.

Zero to the Dealer

A Cannes casino nearly went bankrupt from a baccarat game in 1957.

The Palm Beach Casino was the scene of the richest baccarat game in history when two Hollywood producers won nearly $800,000 over several days of play.

Their opponents were the Greek syndicate that held the bank on behalf of the casino. During the game, chips had to be replenished three times, and when a fourth new bank was called for, the casino announced it was broken.

The game marked the last time the Palm Beach ever held a baccarat bank. But the two Hollywood producers, Daryl F. Zanuck and Jack Warner, found other games.

THERE ARE
20,609,637,968,000,000,000,
000,000,000,000,000,000,000,
000,000,000,000,000,000,000,
000,000,000,000,000,000,000,
000,000 DIFFERENT WAYS OF PLAYING
THE 30 MOVES ON EACH SIDE IN CHESS

Greek Mythology

Jimmy the Greek is a man who doesn't believe in chance.

He's a bookmaker, an oddsmaker, an oracle and a television personality. And he's usually right.

He won his first large bet on the 1940 presidential election when he got the best possible odds on Roosevelt, shortly after organized labor backed Willkie. He followed that up by betting heavily on Truman in 1948 after he discovered women didn't like men with moustaches, a fact that dropped Dewey out of contention in his book.

He was a millionaire at age 30 and he's still going strong. Some recent predictions:
* Ronald Reagan vs Jimmy Carter: 5-2 Reagan (Reagan won)
* John Anderson for president: 25-1 against (right again)
* George Brett batting .400: 5-2 against (Brett didn't make it)
* Larry Holmes vs Muhammad Ali: 11-5 Holmes (Ali didn't make it)

Gambling Tales

Everybody has his favorite gambling story, but chances are you never heard these:
* Pete Rose set a National League batting record in August 1981 by getting his 3,631st hit—and in the same month, variations of those numbers won lotteries in both New Jersey and Delaware.
* 20,000 homing pigeons were entered in the annual Pigeon Fancier's race in Yorkshire, England, in 1972. Only 500 finished the 150-mile course; 19,500 disappeared.
* First prize in the Miss Zambia contest in 1972 was $350 and a cow. Second prize was $175 and a pig.
* Harold Levitt dunked 499 consecutive free throws on April 6, 1935, but he never played college or pro basketball. He was only 5 feet 4.
* Several employees of the Sunrise Hospital in Las Vegas were suspended in 1980 when administrators discovered they were making bets on how long certain patients would take to die.
* A mathematics professor in England worked

out that a player hits the jackpot in a slot machine once in 4,000 times. At a nickel a throw, it would cost you $200 to win $5.
* The seventh earl of Derby bet on the horse races for 38 years and made $3.2 million.
* Charles Macey of Crowbarrer, England, won $1 by making 12,000 consecutive jumps on a pogo stick—more than seven miles.
* An anonymous sailor made a record 27 straight wins with the dice at a crap table in the Desert Inn in Las Vegas (the odds against this happening are 12,467,890 to 1). If he had let his money ride on each throw he would have won $268 million. But he didn't and walked out with only $750.
* You can load dice by letting them stand for a few weeks in a saucer of mineral water. The impregnated side of the dice will be heavier and wind up on the bottom when you throw.
* The first 20 moves in chess can have 169,518,829,100,544,000,000,000,000,000 variations.
* The odds against a dead heat in a horse race are 750 to 1.

The Golden Dream

Suddenly—You're a Millionaire!

It's the affordable dream, the stuff of which fantasies are made. Yet you've got a better chance of being struck by lightning (2,000,000 to 1), killed in a hurricane (2,500,000 to 1) or dying in a nuclear reactor accident than winning many lotteries or scooping the big one in Britain's football pools.

In a recent Pennsylvania instant game, for example, the odds of winning a million were 30,000,000 to 1. And the odds against winning New York's $1 state lottery of $1million are 25,000,000 to 1.

Backing Britain

Britain has over 100 lotteries with ticket sales in excess of £200 million. And 17 million football-pool coupons are sold each week—that's 35 percent of the nation's adult population!

THE NUMBER 8386 TWICE WON $20,000 IN THE MEXICAN NATIONAL LOTTERY NOV. 4, 1938 and AUG. 4, 1939

The Fat One

Spain's Christmas lottery is called El Gordo, the Fat One, and pays out more prize money than any other lottery in the world.

In December 1979 it paid out over £200 million—£40 million to members of one church in the small town of Branollers. The priest had sold tickets on condition that one quarter of the winnings go to mend the church roof. The church won £10 million!

The Big Bite

Miron Vieira de Sousa of Brazil won £1,065,891 for a 25-pence bet on the Brazilian football pools in 1975. The first thing he did? He went out and bought himself a set of false teeth!

Luck of the Irish

The Irish Sweepstakes sells tickets in 147 countries in aid of medical services in 98 countries. It holds six draws a year with well over a million tickets sold for each one. Odds of winning if 1,125,000 tickets are sold? 135,000 to 1.

Voltaire Figured It Out

Voltaire made his fortune—by being good at math.

The 18th-century writer and philosopher quickly realized that French government officials had miscalculated when they set up a national lottery. He put together a syndicate and bought up every ticket. His take made him financially independent and enabled him to devote himself to writing.

A National Disaster

England's first national lottery was a disaster—for all concerned.

Queen Elizabeth I announced in August 1567 that a national lottery would be held to raise money for her depleted coffers. There were to be 400,000 tickets costing 10 shillings each, and of the £200,000 raised, £100,000 would go to prizes. Top prize was a princely £3,000 to £5,000 in cash, with the rest in plate, hangings, tapestry covers and linen.

But tickets sales were appalling—the odds against winning one of the twelve £100 prizes were an enormous 16,000 to 1. The draw was postponed several times. Criminals coming to town to buy tickets were promised freedom from arrest to spur sales.

Nothing worked, and since only one twelfth of the anticipated £200,000 had come in, prize money was reduced to £9,000. The draw finally started on January 11, 1569, and dragged on until May 6—because of the number of blank tickets. Even the queen's own tickets failed to win!

Ivy League Flutter

Lotteries helped build several American universities: Harvard, Yale, Dartmouth, King's College (now Columbia), William and Mary, Union and Brown.

Lots of Lotteries

In 1833 there were more than 420 lottery offices on the eastern seaboard of the United States, more than 200 in Philadelphia alone! Prizes topped $66 million—more than six times the national budget.

George Washington and Thomas Jefferson both favored lotteries, and Jefferson applied for a license to hold one himself in 1826.

Riches in Store

Betty Rawlins and Elenore Ratkowski met and chatted briefly in a town store in western New York in 1973. Betty bought a lottery ticket in the checkout line ahead of Elenore—and won $100,000. Almost five years later, Elenore won $114,000—with a ticket bought at the same store.

God Moves in a Mysterious Way...

Peter Pulaski went to church to pray for $5,000. He won $100,000 in the New York State lottery and told God: "You kind of overestimated my need."

34,000,000,000,000 to 1!

Marty Timmons qualified for the final draw in the New Jersey lottery in 1976 and won $10,000. An hour later he won $1 million with the same ticket. Odds against this? More than 34,000,000,000,000 to 1. It all happened because the ticket number was unusual: 22. 34. 22. The agent had tried to sell it throughout Marty's plant. The day before the draw, bubble-gum cards left in his wife's purse said: "Something of great importance will happen to you" and "You will soon inherit a vast fortune."

Tempting Fate

Mr. and Mrs. Tom Drake of Pennsylvania spent $14,100 on lottery tickets in 1977. They won $15,000 but lost more than $6,000 in wages. They told the press their story, however, and earned several thousand back.

The House That Luck Built—but Abandoned!

"Heartbreak House" in Lincoln, Rhode Island, was built by Stephen Smith after the girl he loved said she would only marry him if he gave her exactly the kind of home she wanted. He won the $40,000 to pay for it in a lottery but the girl never married him—and Smith never lived there.

Three in a Row!

The Lions Club of Nevada, Iowa, held a draw for door prizes in 1962. Rev. James Dendler, Henry Scudder and William Dial were called on to make the draw—and each came up with his own number!

The Professor's Gamble

Prof. Francisco Garrido announced in a lecture in Madrid, Spain, that he was resigning to buy a popular pharmacy with his winnings in a national lottery—the draw for which had not yet been held. His ticket won—and he became the wealthiest pharmacist in the country!

Lucky Row!

Occupants of three adjoining buildings in Queen Street, Woolahara, Australia, won three first prizes of $25,000 each in the state lottery—within nine months. The lottery takes in an area that is larger than the United States and that has a population of more than seven million people.

Fascinating Facts

∗ Five U.S. millionaire lottery winners in 1968 didn't bother to claim their tickets.
∗ Three millionaires had bought only one or two tickets before in their lives.

Choosing Your Words Carefully

Ever wonder why your carefully chosen slogan of "25 words or less" fails to win that dream vacation in Hawaii? Just the way things fall out? The problem could be that you're too clever!

"You have to write simply and stupidly to win," says William Sunners. And he should know. He's been entering contests since 1921 and his winnings include: diamond rings, three cars, six TV sets, five watches, 10 radios, two washing machines, three toasters, two complete wardrobes and 960 cans of tuna. And bundles of cash!

Cashing In

The Case de Cordero in Madrid, Spain, was acquired by Santiago Alonso Cordero in 1860 when he won the grand prize in a national lottery. The Spanish government didn't have the $900,000 in cash which Cordero won as his prize, so he demanded and got the building in the center of Madrid.

Private Fortune

Gaston Doumergue (1863–1937) amassed no wealth while serving as president of France from 1924 to 1931—but bought a lottery ticket on returning to private life and won $100,000.

Math Error Wins a Fortune

English novelist Mary Russell Mitford (1787–1855) was 10 years old when she dreamed of the number 7 on three successive nights. So she multiplied 7 x 3, and bought a lottery ticket on the number 22. Her math error won her £50,000!

The Same Old Car

Dennis L. Wheat of Malvern, Arkansas, won a used car in a raffle in 1963—then discovered it was the same car he'd traded in six years earlier.

The Lucky Ones

Fortune smiles on some people: they win lotteries. But are they really that lucky?

Nearly all winners say loneliness is a big problem. Many leave their jobs, try to conceal their win from relatives, friends and "well-wishers" who harass them with offers.

Many receive less than half the amount advertised as the prize and lots feel they're being persecuted by—guess who?—the tax man! Vicious gossip is also a problem, though it dies down eventually. Most U.S. lottery winners lead "normal" working-class lives, surveys reveal. They have a variety of jobs at low wages.

But in a survey conducted by Dr. Roy Kaplan, most winners wound up feeling more secure. Few changed their way of life very much; most watched more TV, drove more, worked around the house and read the newspapers. Few spent their winnings on culture but many traveled to faraway places and on round-the-world trips. Most gave to charity.

New World Gamble

The 13 original colonies in America were largely financed by lotteries in England—the Virginia Co. held one in 1612 to help pay for the settlement of Jamestown.

In 1776, the Continental Congress voted to hold a lottery to raise $10 million for the American Revolution, but the scheme was abandoned.

With a Little Bit of Luck

Guards escorting the French government Lottery Wheel.

Drawing a lottery in the Guildhall, London, 1739. During the 18th century this method of raising money was often used, in this instance to provide sums for building a bridge at Westminster.

Practicing What She Preaches

Entering contests certainly changed Selma Glasser's life. She learned to drive only after winning a car. Selma, who runs a course at Brooklyn College teaching people how to win competitions, has also won: vacations to Europe, the Caribbean, the Catskills and a Texas dude ranch; a freezer, a mink, a gas oven, a home-heating system and a motorcycle. And dates with Sid Caesar and Engelbert Humperdinck!

Positive Thinker

When Helen Hadsell decided she wanted a new home, she won one. She just knew she was going to win and had to act surprised when contacted by the organizers! She's also won: a fridge, a microwave oven, furniture, TV sets, tape recorders, cameras, toys, bicycles, a car, cash—and numerous vacations.

Getting a Word in Edgeways

The odds of winning a slogan competition? That depends on the number of entries, prizes and correct replies. In the U.S. a major national promotion may draw eight million entries; more than 30 million readers have been known to enter contests run by national magazines.

Best bets: local contests and unadvertised giveaways.

Winning Wins

Here are some tips from the experts on how to better your chances:
* Use as many of those 25 words as you can—and work on them.
* Try quips and plays on words. Use aids such as dictionaries, thesauruses and catalogs.
* Follow the rules, and study the judges and the product's promotional literature.
* Try a bit harder at Christmas and in the summer. There's more competition then.

THE **BET** THAT PROVED PROPHETIC!

John SCOTT (1751 - 1838) RELUCTANTLY MOVED HIS LAW PRACTICE FROM NEWCASTLE, ENGLAND, TO LONDON AFTER SHOWING HIS LACK OF FAITH IN THE GREATER OPPORTUNITIES THE CAPITAL OFFERED BY GIVING ODDS OF **1,000** TO **ONE** ON A BET THAT HE WOULD BECOME **Lord High Chancellor!**

25 YEARS LATER SCOTT PAID THE BET OF **$5,000** WHEN HE BECAME LORD ELDON AND WAS GIVEN THE HIGHEST JUDICIAL OFFICE IN ALL ENGLAND

WILLIAM CROCKFORD A 19th CENTURY BRITISH GAMBLER, DIED BEFORE HE COULD COLLECT HIS WINNINGS ON THE DERBY OF 1844 — SO CRONIES *PROPPED HIS BODY UP IN AN ARMCHAIR IN HIS WINDOW UNTIL ALL HIS BETS HAD BEEN PAID*

MRS. **MINNIE GAWELL** of Chicago, Ill. DRAWING IN A RAFFLE PICKED HER OWN NUMBERS **4 TIMES IN SUCCESSION**

Sports

*T*he Big Game is an irresistible lure for the inveterate gambler. He covers all the angles, weighs all the odds—and still draws a blank when fate steps in.

Chance certainly works in strange ways when it comes to sports. The best man or team often doesn't win—be they golfers, footballers, soccer players or hockey players.

Here are some random examples of chance in action on the sports field:

* The winning baseball manager who always made sure his team saw a beer truck before a big game—for luck.
* The man stabbed to death—by a baseball!
* The baseball players who like to wear number 13 and who refuse to change their lucky clothes.
* The strange superstitions of the great soccer player Pele, of the world's greatest hockey players and a former world heavyweight boxing champ who is scared of black cats.
* The hole in one scored by an earth tremor, and the golfer who drove a ball into a man's pocket, 250 yards away.

Read on and ponder the ups and downs of the world of sport, where chance is the key member of the team!

Blue Jays Baseball team at Exhibition Stadium, Toronto.

Good Sports

The best bet in sports? Professional football, say the experts. But to load the dice even more in your favor, look for the following when calculating your bet:
* Team injuries.
* The weather.
* Home-field advantage.
* Head-to-head records.

The same applies to most other sports:
* Basketball: Look for home-court advantage and point spreads, particularly late in the game.
* Hockey: Consider home-ice advantage and the goalies' records, along with scheduling.
* Baseball: Study pitchers' records, injuries and winning streaks. And don't bet until June, when the true mettle of a team begins to show.

And one final tip:
* Don't bet the wire-service ratings. They're the worst possible guide!

How You Play the Game

Sports is big business. And with millions of dollars riding on a strong arm or a healthy knee, it's hardly surprising that many sportsmen woo Lady Luck with elaborate rituals. Baseball players believe:
* It's unlucky to step on foul lines.
* You can beat another team by putting a black cloth in their dugout.
* You must never remind a pitcher he's got a no-hitter going.

John McGraw, manager of the famed New York Giants for 31 years, always told his players to look for a beer truck before the game. It was lucky, he said.

He himself didn't leave things to chance, however. He hired a beer truck to pass in front of the park before every game of one World Series. It worked. The Giants won.

The Big Pitch

Fourteen-year-old William Cartwright of Harlingen, Texas, pitched four consecutive no-hit baseball games for the Harlingen Colts in 1953. Odds against pitching one no-hitter: 1,300 to 1.

Percentage Plays

The average baseball team is beating itself by not making the smart plays, says computer expert Earnshaw Cook. And he reckons any team can add 273 runs to its total and rise from last to first place by following his advice, gleaned from 750,000 statistics from 12,000 games.

He believes:
* Starting pitchers shouldn't start. Managers should save them for the middle three innings and use relievers the rest of the time.
* Pitchers should never bat for themselves unless the team has a comfortable lead.
* Players should hit in order of ability.
* The sacrifice bunt should be scrapped and hit-and-run plays should be used more often.

It's been estimated that Cook's recipe for success could mean the difference between a losing and a winning season in the audit books, adding about $4 million a year to gate receipts.

Oh, and sports addicts might like to know that Cook thinks the average baseball season is about five times too short to make predictions about batting averages!

Lucky Strikes

Batters and pitchers go through an elaborate ritual of cap-touching and foot-shuffling before they play ball. Here are some other players' superstitions:
* Several like to wear number 13, including Dave Concepcion of the Reds, and Roy Howell, who asked the Toronto Blue Jays for it. Pitcher Pete Vukovich thinks Friday the 13th is his lucky day.
* Veteran pitcher Lois Tiant is called The Man from Glad. Why? Like the man in the TV commercials, he wore only white throughout one season—for religious reasons.
* Leo Durocher refused to change his clothes all through the Giants' 1951 pennant drive. And Walt Alston always wore the same shirt through his club's winning streaks.

Going for Broke

Detroit Tiger Rudy York hit two homers over the fence in a game in 1937. Both broke windows—in the same car!

Chico Carrasquel hits a home run.

Record-breaker Ty Cobb sliding into 3rd base.

Flied Out

John Lubinski hit a pop fly in a game in Minneapolis in 1927 but fell while running to first base. The ball landed in his own pocket!

Stanton Walker was sitting between two friends at a baseball game in Morristown, Ohio, in 1902. One friend was passing an open pocket knife to the other when it was hit by a foul ball just as it was passing in front of Walker. The impact drove the knife into Walker's heart, killing him.

The Olympics

No sporting event in the world is bigger than the Olympics, but some stories of winners— and losers—defy chance:

* The hero of the 1904 marathon at the St. Louis Olympics was a Havana postman named Felix Carvajal who wasn't even chosen for the Cuban team. He earned expense money by jogging around the Havana civic square and collecting coins from the crowd; he lost all his money in a dice game in New Orleans and had to hitchhike to the race; he showed up in heavy boots and had to borrow a pair of sneakers; during the race he liked to stop and chat and used to detour through orchards to steal apples. Did he win? No, but he was fourth!

* Wyndham Halswelle of Great Britain became the only athlete in history to win a gold medal by running a race by himself. Officials at the 1908 Olympics ruled that Halswelle had been forced off the track by two Americans—J. C. Carpenter and W. C. Robbins—who beat him. They ordered a rerun, but the Americans walked out and Halswelle jogged by himself.

Some Odds

Chances are you didn't know:

* Once a player signs with a baseball club, the chances are one in 20 he'll make the majors, one in 1,500 he'll reach the Baseball Hall of Fame and one in 23,000 he'll win the batting title.

* The probability that the team favored by Las Vegas bookies at the start of the World Series will actually win it is 44 in 100.

* The home team wins in the National Basketball Association 63.4 times out of 100.

* In a title bout, the odds against the challenger dethroning the champ are 4 to 1 (heavyweight), 5 to 1 (lightweight) and 2 to 1 (middleweight).

* The convert in American football is successful 903 times out of 1,000.

* If you're into horses, the chances are 41 in 100 a thoroughbred will earn its keep.

* The chances of winning Wimbledon in straight sets in the final are 62 in 100. But the chances are 82 in 100 that the player who wins the first set will win the match.

The Right Stuff

Satchel Paige, the oldest man in baseball and one of its most famous pitchers, didn't break into the big leagues until 1949 when he was said to be 42 at least. He was elected to the Baseball Hall of Fame in 1971 and he credited his success to these rules:

* Avoid fried meat.
* If your stomach is upset, lie down and quiet it with cool thoughts.
* Keep the juices flowing by moving gently.
* Go light on vice and don't carry on in society.
* Avoid running.
* Never look back.

$ports

Chances are you never knew this about sports.
* Every year about 300 million tickets are bought for major sporting events in the United States and the gross revenue, including admissions, parking and concessions, is $1.8 billion.
* TV sales bring in another $1 billion to teams.
* Direct participation in sports involves an estimated 286 million Americans and produces $10 billion yearly, mainly on equipment and travel.
* The annual recreational expenditure in the United States is $30 billion.

Believe in Chance?

We can all be superstitious, but some of the world's greatest golfers take their beliefs beyond the laws of chance:

* Chi Chi Rodriguez refuses to use a ball with the number 3 on it and always uses the head side of a coin to mark his ball. He also likes to wear green on Sundays because it is the color of money.

* Hubert Green refuses to wear yellow because it's "passive."

* Al Geiberger marks his ball with a penny and, if he's having a good round, puts the coin in the same place beside the ball each time. He also likes to line up the eyes of Abraham Lincoln on the coin with the hole.

* Tom Watson will eat the same thing for dinner he had the night before a good round. He also carries three coins and three tees in his right pocket. And, he always uses a broken tee on par-3 holes.

JIMMY **NICHOLS**
ONE-ARM
GOLFING STAR
MADE A
336-YARD
HOLE-IN
ONE

Douglas,
Georgia
1933

ALTHOUGH LEFT-HANDED -
HE USES A RIGHT-HAND STANCE

JAMES EAST
San Diego
Calif.

HAS MADE
65
HOLES-IN-ONE

———

60 WERE
MADE ON
PRESIDIO
HILLS COURSE
1490 YDS.
LONG

Chi Chi Rodriguez: double bogey is no laugh.

Gary Player, the South African golf veteran, gives his gallery some laughs while conducting a golf clinic. He'll opt for black attire in serious competition 'tho.

Tom Watson pops out of the reeds.

Ice Capades

Although the game is not as popular as baseball and football in North America, a lot of people still bet on the National Hockey League. Bookmakers say they keep these things in mind when accepting a bet:

＊The home rink is a major advantage in hockey and home teams win a larger percentage of games in hockey than either football or baseball.

＊A goaltender's record is the single most important factor in making a bet.

＊The tough schedule many hockey clubs keep is another factor. You can usually always bet against the visiting club in the second or third game of a road trip—no matter how easy the opposition.

＊Also go with the team that has a good power play—especially when it's playing a team that draws a lot of penalties.

Boston Bruin goalie Gerry Cheevers kept his mask in stitches for 19 seasons.

The greatest defenseman of all time: Bobby Orr.

Phil Esposito wheels away from Russia's Vladimir Petrov.

Hockey

A Bobby Orr or Wayne Gretzky can work magic on the ice. Other hockey players have to invoke the gods of good fortune. Most circle the goal and tap the goalie for luck before a game. And no one must see the Stanley Cup before it is safely won. The custodian of the cup has to stand in front of it in the final game, even when he wheels it out with two minutes to go.

Then there's:

* Gary Dornhoefer, who discovered that he'd been wearing his contact lenses in the wrong eyes. But he refused to change.
* Gordie Howe, who always borrowed a teammate's stick when he was on a hot streak. He even borrowed them from other teams.
* Bobby Orr, who most of the time touched everybody on the ice before a game.
* Phil Esposito, who wore a black shirt backwards for every game. The reason: he'd once scored three goals similarly clad!
* New York Rangers goalie Gilles Gratton, who believes he once stoned people in another life and now they're getting even.
* Indianapolis Racers goalie Andy Brown, who plays without a mask and has a "lucky fan."
* Gary Cheevers, who has his mask painted with fake stitches—to ward off the real ones.

Wayne Gretzky, whiz kid from the Edmonton Oilers, moves in for the kill.

Luck to the Fore

My Honor

There's chance and luck and fate, but this one is ridiculous.

James Cash was playing golf on a course in Belmont, Massachusetts, in 1929, when he drove a ball from the tee to the edge of the cup on a par-3 hole. As he started down the fairway to putt out, an earth tremor hit the course and the ball dropped in for a hole in one.

Like Father, like Son

Chas. H. Calhoun, Sr., and his son Charles E. Calhoun, Jr., both made a hole in one on Washington Golf Course in 1932—within minutes of each other!

Thanks, Partner

Robert T. Jones shot a hole in one on a Long Island golf course in 1951—though his ball stopped 12 feet from the cup. The next player's shot knocked Jones's ball into the hole.

Play It Again, Sam

Pro golfer Sam Snead drove a ball into the pocket of a man 250 yards away!

On His Knees

Dave Ragaini, playing golf on his knees to win a bet, swung a 3-wood on a 207-yard par-3 hole at Wykagkl County Club, New Rochelle, New York—and scored a hole in one.

Charlie Boswell, blind golfer, emerges from the bunker.

Golf fan Robert L. Ripley (left) watches blind golfer Dr. Oxenham drop the last putt. Oxenham had never played golf until he lost his sight.

Hole in One

The odds against making a hole in one in one round are 43,000 to 1.

Black and White

Jack Nicklaus always wears white shoes. And fellow golfer Gary Player always wears black.

One, Two, Three

O. B. Larson of Kansas and his two sons each got a hole in one on the same hole.

Fame but No Fortune

Lloyd Foree, with 20¢ in his pocket, hitchhiked 120 miles to enter a hole-in-one tournament in 1935. He won—scoring an ace—and received a set of golf clubs, but arrived home broke.

Golf champion Jack Nicklaus powers another drive for yet another title.

Golfing great Sam Snead.

Lucky Strikes

One in 10,000

Blind bowler Ben Pearlman of Philadelphia made the 6-7-10 split in 1947. It's one of the most difficult shots in bowling and is only accomplished about once every 10,000 tries by sighted bowlers.

How Not to Do It

Conrad Wild of Milwaukee, Wisconsin, tried to show his bowling instructor the hook that was ruining his game. He rolled 12 strikes in succession for his first 300-game!

Strike!

The odds against bowling a perfect game are 4,000 to 1.

Blind bowler Irene Hallman from Buffalo gets her bearings by special guide rails set up beside the alley.

Perfect Day

Len Wittman bowled seven 300-games in one day in 1915. He made 33 consecutive strikes.

Pinup

Vic Vizvary of Bridgeport, Connecticut, saw one of his pins fly high in the air and then balance upright on top of another pin!

Free Kick

Pawns of Fate

Murar Rao, commander of the Mahratta armies of India from 1740 to 1763, took thousands of prisoners in his campaigns and played chess with each one. If the captive won, he was set free. If he lost, he was hurled to his death from the battlements of Murar's fortress!

Soccer Tales

What are the chances of this happening?

David Grant and Brian Strutt were both in the same school class, on the same boys' teams and both worked as ball boys for English soccer team Sheffield United—and both turned professional with Sheffield.

What a Way to Go

Albert Juliussen, a high-scoring center-forward for Dundee in the late 1940s, once scored 11 goals and finished on the losing side.

It happened during the war when he played the first half for one battalion side, scoring six goals, and then was swapped to the other side and scored five.

Windy Win?

Ray Cashley was playing goal for Bristol City against Hull City in 1973 when he cleared the ball.

It was assisted by a strong wind and bounced over the head of visiting goalkeeper Jeff Wealands and into the Hull net! Bristol won 3-1.

Here Are the Odds

What are the odds against a team scoring more than one goal in a soccer match?

How to Win the Pools

We've heard a lot about taking a chance, but this is ridiculous.

One man told the *News of the World* in Britain that he wins soccer pools by filling his coupon out and then laying it on his breakfast plate. He then sprinkles nutmeg over it and waits exactly 20 hours before mailing it.

Wake Me When It's Over

Fullback John Oakes was playing for Port Vale on December 26, 1932, when the game against Charlton was called because of bad light.

The second-division match was arranged for later in the season and by then Oakes had been transferred—to Charlton. He thus holds the distinction of being the only player to appear for both sides in an English league.

Soccer star Jimmy Greaves scrambles for position.

	2 or more goals	3 or more goals
World Cup	7-6	3-1
European Championship	5-3	5-1
English First Division	12-7	14-3
Scottish First Division	8-5	13-3

The Goal He Didn't Want

Dennis Evans was playing for Arsenal in 1955 in a match against Blackpool when he heard the whistle and assumed time was up.

He booted the ball into his own net in frustration. But the whistle was blown by someone in the crowd and Evans scored on his own team. Arsenal still held on to win it 4-1.

Partners

Chelsea beat Leicester 3-1 in 1954, but the game is remembered for one of the strangest goals on record.

A shot bounced off the crossbar and fell between two Leicester defenders, Stan Milburn and Jack Froggat, who both lunged at it and connected with each other. The ball flew into the net and was recorded officially as "Milburn and Froggat shared own goal against own team."

Four days before the 1970 World Cup qualifying match between Honduras and El Salvador, diplomatic relations between the two countries were severed. The result of the game was a 3-2 victory for El Salvador, preceding a war with 3,500 casualties.

A Foggy Day...

The field was full during a wartime match between Chelsea and Charlton at Stamford Bridge, but only one player was left at the finish.

Heavy fog rolled in and the game was abandoned. But Charlton goalkeeper Sam Bartram didn't know the other 21 players had left the field until a policeman loomed out of the gloom 10 minutes later and told him the match was over....

A Perfect 10

No, it's not Bo Derek. Pele, the world's most famous soccer player, is obsessed with the number 10.

He thinks it's his lucky number and has got it everywhere. Why? He drew the number for both the World Cups he won with Brazil, against odds of 100 to 1.

His real name is Nascimento, which has 10 letters. He always books hotel-room number 1010. And the numbers on his second car add up to—you guessed it—10.

Champion Jomo Sono tackles well, but fans on his shot.

The one and only Pele.

Laced with Luck

The National Football League

NFL betting is the most popular pastime in America, but it's illegal in most states. However, nothing stops the mad weekend rush to beat the spread and predict the winner by giving or taking points. What are your chances of winning?

Here are some pro tips:

* Gamblers say the biggest winning edge a bettor can develop is knowing the full extent of injuries to players.
* Weather is also very important. Some southern teams like Los Angeles, Dallas or Miami do not play well late in the season in cold-weather areas like Minnesota, Green Bay or Detroit.
* Home-field advantage and whether the team has played on a Monday instead of a Sunday the previous night. Successful bettors keep quarterback records comparing home versus away and passing percentage. They also use the adage that Monday night is worth a field goal—three points—when a favorite is playing a fresh underdog.
* Chances of becoming a pro player: of one million playing high school football, 21,500 will get college scholarships and only 333 will make it to the National Football League.

Montreal Alouette fullback John O'Leary is head over heels as he is blocked on a short gain.

Over the top: the human wall can't stop this one-yard touchdown dive.

Laced with Luck

Quarterback great Joe Namath ties his shoelaces right over left. Why? "I've always done it, and life has been good. So why change?"

Covering All the Angles

The ancient Greeks and Romans believed the sea, rivers and lakes all were populated by gods and goddesses.

Modern-day fishermen are just as anxious to court the Goddess of Fortune. The wise angler:

* Always throws the first worm into the water when baiting his hook.
* Always lets the first fish go.
* Never uses a net.
* Recites the old rhyme: Fish east, fish bite least; fish west, fish bite best.

Fish Tales

William Bardwell caught the same fish—three times. Fishing in the Welland Canal, in Ontario, Canada, he hooked a 25-pound pike and threw it back. Twenty minutes later, he caught it again...and threw it back. Moving 100 yards upstream, he caught it again.

Hooked

Earl Moore of Forgan, Oklahoma, caught an eight-inch trout—without his hook touching the fish. The trout had been hooked some months before and the hook remained in its mouth. Moore's hook went through the eye of the other hook.

New York Jets Super Star Joe Namath blows bubbles and hopes for another touchdown.

Some Bite

James Price lost his dentures in Bull Shoals Lake, Arkansas—and got them back 10 days later when he caught a 20-pound catfish that had swallowed them.

Ring of Truth

Howard Ramage's wife lost her wedding ring down a drain in 1918. A Vancouver man found it 36 years later—in the stomach of a fish—and returned it to Mr. Ramage.

Lucky Punches!

﹡ Boxer Danny London was born deaf and dumb. But after he was hit on the head during a fight in Brooklyn, New York, in 1929, he suddenly found he could both speak and hear!

﹡ Muhammad Ali became heavyweight champion of the world by chance! When he was a boy his bicycle was stolen—and he took up boxing so he could clobber the thief!

﹡ Heavyweight boxer Ken Norton fears no man in the ring. But he is afraid…of black cats!

﹡ The odds against the challenger winning a world heavyweight championship? 4 to 1.

Injuries

What are the chances of an injury snuffing out a career or—even stranger—spurring an athlete on to greater heights? Judge for yourself:

﹡ Babe Ruth got a bellyache during the spring of 1925 and had an operation. He returned

America's favorite; Joe Di Maggio.

The greatest: Muhammad Ali

to the lineup June 1, but never recovered, playing in only 98 games and hitting only 25 homers. The Yankees were also sick, dropping to seventh place.

* Quarterback George Shaw was injured while leading the Baltimore Colts against the Chicago Bears in 1966. He was replaced by a youngster who had been cut by the Pittsburgh Steelers. And the kid was good. Johnny Unitas became one of the greatest quarterbacks in history.

* A bone spur caused Joe DiMaggio to miss the first 65 games of the 1949 season. But when he returned he slugged four homers and drove in nine runs. He eventually played in 76 games, batting .346, and led the Yankees to the world championship.

* A pitcher named Stan Musial was playing for the Daytona Beach Islanders in 1940 when he injured his left shoulder. He concentrated then on hitting and won seven National League batting titles.

* Boxer Buster Mathis was picked to represent the United States as a heavyweight in the 1964 Olympics, but a training injury forced the team to go with the alternate. His name?

Joe Frazier. He won the title and later the heavyweight championship of the world.

* Golfer Ben Hogan was nearly killed in 1949 in an auto crash, but he returned 10 months later and won the United States Open for the second time.

On the Ball!

Otto Reselt made a perfect six-cushion pool shot with 120 balls on the table, the cue ball hitting the cushion six times and traveling more than 40 feet in its journey round the table, missing all the balls except the two aimed at.

The Most Famous Marksman of Them All!

Andrew Jackson (1767-1845), the general who became the seventh president of the United States, was a crack shot. Firing a new pistol for the first time, he hit a target consisting only of two crossed threads three times in a row. And he was 40 yards away from the target!

The Immortal Babe Ruth slugs another one out of the park.

PART THREE:

Rich Man, Poor Man

Free Enterprise and the Lucky Break

*I*f chance has a partner, it is money. The two walk hand in hand through history, aiding and abetting each other, laughing at one person, but blessing another. Fortunes have been made because chance chose to take one person and make him—or her—a millionaire.

In this chapter, lives, fortunes, family, life and death are ruled by chance. We learn of the 18-year-old girl who discovered a huge ruby, and of the man who found gold because he had been drinking.

There are stories about people's obsessions with sunken treasures, a fisherman's catch that opened a treasure trove, the search for the *Titanic*.

You will read about the most eligible men and women in the world, about a ruined man who turned the Depression into a fortune, about a tentmaker who made millions.

There's a fat housewife who shared a secret to make her fortune, a man who needed Perry Como, and two countries that built an Edsel of the air. Meet a millionaire who built a house with two views, learn about the deficit in the Vatican and, most important, find out how YOU can make a million.

Prospectors ascending the summit of the Chilkoot
Pass in 1898 during the Klondike gold rush.

Buried Treasure

We all dream of striking it rich, but some people are so lucky they just dig it up. The following stories defy the laws of chance:

* Kim Jones, 18, was looking for interesting rocks at an old mine near her home in Franklin, North Carolina. She found a 456-carat ruby valued in 1978 at $15,000.

* Police in Bethlehem, Pennsylvania, were wondering who was flushing $100 bills down the toilet when sewer workers found more than $4,000 in 1977.

* Gold coins worth $24,000 were found in an old shack in Lovelock, Nevada, when it was bulldozed for new construction.

* House painter Danny Diaz found $50,000 in an old house he was working on in California.

* Rev. Stanley Jones was cleaning his parish church near Cheshire, England, in 1969, when he found an old chest. It contained antique silverware dating back to 1677 and worth more than £30,000.

* Bulldozer driver Malcolm Tricker was razing a building in Suffolk, England, when he found five gold collars dating back to 100 B.C. He turned his find over to the British Museum for £20,000.

* Another construction worker in England found more than 1,000 gold coins four feet underground near Newstead Abbey, once the home of Lord Byron.

THE SADDEST SWINDLER IN HISTORY
JACK GILLIN "SALTED" THE APPARENTLY WORTHLESS WYOMING MINE IN HOMANSVILLE, UTAH, BY EXPLODING INTO ITS WALLS HIGH-GRADE ORE - THEN SOLD IT TO UNSUSPECTING PURCHASERS FOR $22,000. THE MINE SHAFT WAS EXTENDED 400 FEET FARTHER AND THE NEW OWNERS STRUCK A FABULOUS BONANZA OF GOLD AND HIGH-GRADE SILVER THAT MADE THEM MULTIMILLIONAIRES (1873)

Diamonds Are Forever

A poor farmer discovered the world's richest diamond mine in 1866, but had trouble convincing people that diamonds could be found in South Africa.

Schalk Van Nierek noticed an odd-looking stone among pebbles near Hopetown, South Africa, and bought it from a group of children who were using it in a game.

He tried to interest people in buying it, but got no takers. He then took it to a

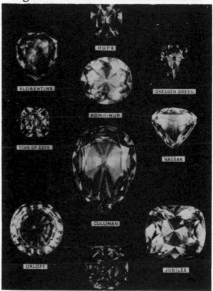

Famous diamonds.

doctor who vowed it was a 21¼-carat diamond. It was sent to England and authenticated, but still people refused to believe the story, suspecting it was a fraud.

Sir Robert Murchison, the highest geological authority in Britain, staked his reputation on the idea it could not have come from South Africa. Minerologists agreed, speculating the stone was eaten by an ostrich and later dropped in its dung.

Van Nierek persisted and teamed up with a local trader to locate more gems. Finally, the two produced a huge 83½-carat stone and people finally began to believe that diamonds existed in South Africa.

The hunt turned into a tide and Van Nierek's huge stone was called the Star of Africa. The country still is the world's best source of diamonds.

Hard-Luck Charlie

Charlie Steen had enough misfortune to discourage any man.

In 1952 he was stone broke, standing on a peak in Utah with his broken drill in pieces. He was living in a $14-a-month shack, his four children were starving and he was prospecting for uranium even though he didn't have money for a Geiger counter.

By chance, Hard-Luck Charlie looked at his broken drill bit and noticed a strange blue-black substance. He took it to a garage and borrowed a Geiger counter. The instrument went off the dial and never wavered.

Hard-Luck Charlie had discovered pitchblende, the world's richest source of uranium. He borrowed a dollar, filed a claim, and his find was eventually worth $60 million.

But even the fortune didn't change his luck. Within 15 years Charlie had spent every dime on parties, palaces, yachts, antiques, women and liquor. By 1968 the tax department had taken the rest and Steen went back to prospecting, still wandering around the West.

The Oakes Mine

A down-at-the-heel prospector discovered gold in Canada because he was kicked off a train for drunkenness.

The saga of Harry Oakes defies chance. He spent his life searching for gold and arrived too late for the big strikes in California, Alaska, the Belgian Congo and Australia.

In 1900 he landed in Canada where he was thrown off a train near Kirkland Lake in Ontario. He bought a claim for $5 from a starving Chinese cook and then added another from a crazed prospector who asked only for a crust of sourdough and a plate of rancid beans.

Finally, after months of backbreaking work, Oakes clutched a high-grade nugget in his hand and founded the Oakes mine.

By 1921 he was one of the wealthiest men in the world. He turned his back on Canada and bought himself an English knighthood, retiring to a mansion in the Bahamas. He was murdered there in 1943, and the crime was never solved because there were too many suspects.

STRUCK OIL ACCIDENTALLY WHILE EXHIBITING A DRILLING SET!

PORTABLE ROTARY DRILLING UNIT SET UP FOR DEMONSTRATION PURPOSES STRUCK OIL AT THE INTERNATIONAL PETROLEUM EXPOSITION IN TULSA, OKLA

Raise the *Titanic*

Searchers attempting to find the *Titanic* are still waiting for chance to step in.

The ill-fated liner smashed into an iceberg on its maiden voyage on April 15, 1912, and sank in 12,000 feet of water southeast of Cape Race, Newfoundland, killing more than 1,500.

Since then, numerous attempts have been made to locate the famous ship, but experts fear that the winds and currents in the area may mean it is scattered in pieces on the Atlantic's floor.

There have been discoveries of what appear to be bits of the ship, and there was a multimillion-dollar search using small submarines, but the old liner is still in its watery grave.

The prize for finding it is big. It was reportedly carrying more than $300 million in diamonds and other valuables.

Caribbean Salvage

Melvin Fisher took a chance, but he nearly lost the gamble.

Convinced he had located the grave of a Spanish galleon in the Caribbean, Fisher decided to devote a year to finding it. But, with only five days to go, all he had for his troubles were fish.

Then, incredibly, his underwater vacuum system started coughing up Spanish doubloons. He brought 2,000 of the gold coins to the surface—one was sold recently for $25,000.

But Fisher's good fortune didn't last. While he was searching for two other ships his son and daughter-in-law were drowned.

Great Treasure

The Caribbean is not the only treasure trove of ship's booty.

Historians and sea buffs estimate more than $1 billion worth of gold and artifacts lies at the bottoms of the Great Lakes. But no one yet has taken the chance and mounted a major expedition.

A strange marine salvage operation was effected by Japanese fishermen in 1900, a century after a precious cargo of porcelain had sunk to the bottom of the sea. The fishermen attached live octopuses to a line and lowered them to the fragile treasure. The octopuses curled about the vases and they were safely hauled up to the surface. Believe It or Not!

In Xanadu...

An earthenware pot in a fisherman's net may have given us the world's richest sunken treasure.

In 1980 fishermen off the coast of southern Japan began pulling up cups and pots in their nets, and a local professor became interested. Torao Mozai obtained a $60,000 grant and hired divers, who found the hulks of 72 wooden ships on the bottom.

The old ships have been identified as the remains of Kublai Khan's fleet that tried to invade Japan 700 years ago. They contain a treasure trove so rich that each item is being recovered individually.

The personal seal of the great Khan has already been recovered. Mozai believes the ships also contain a 12th-century statue of Buddha and thousands of priceless examples of Mongolian weaponry.

The *Andrea Doria* Story

New York millionaire Peter Gimbel was always fascinated by the sinking of the *Andrea Doria*, the Italian luxury liner that went down in 1956 in the Atlantic about 45 miles from Boston.

He financed an expedition to dive on the wreck and locate its bank and safes. Finally, in 1981, his efforts were rewarded and divers, using high-powered underwater vacuums, found the ship's bank and purser's office.

Safes estimated to contain between $1 million and $4 million were hoisted to the surface, uncovering the treasure of the *Andrea Doria*.

Your Own Backyard

If oceans aren't your style, how about the backyard or garden?

The Federal Reserve Bank in the United States estimates that $50 billion worth of treasure has been buried across North America by robbers, tax evaders and people who hide their money and then forget it.

Chance discoveries are reported all the time, and in the U.S. if there is no surviving claim, the money is yours.

Something for a Rainy Day

A chance walk in her backyard proved lucky for a widow in Rhode Island.

Shortly after her husband died in Woonsocket in 1955, the widow noticed he had built a secret chamber six feet wide under a shed in the yard.

She dug it up and found more than two tons of silver coins!

© King Features Syndicate, Inc., 1962. World rights reserved.

THE BONANZA THAT BROUGHT ONLY DEATH !
Marie Countess Arco of Austria
FOUND $50,000 IN GOLD DUCATS IN A CHEST IN HER GARDEN
—BUT NEVER SPENT A SINGLE COIN
SHE CARRIED IT STRAPPED TO THE LUGGAGE RACK
INSIDE HER COACH EVERYWHERE SHE WENT
UNTIL ON JUNE 23, 1848, A BUMP DISLODGED THE
TREASURE CHEST AND IT KILLED HER !

Bright Ideas

Chance could smile on you if you're inventive, bright and won't take no for an answer. How often have we said to ourselves, "If only I'd thought of that." Several people came up with that one big idea first and parlayed their beliefs into a fortune. Here are some of their stories.

Look It Up

A salesman who didn't like his first name started the world's first catalog store.

Aaron Montgomery Ward was just another drummer (salesman) traveling around America's Midwest in the 1860s when chance stepped in. Ward noticed many of his customers were angry at paying commissions to middlemen, and he wondered if he could sell his goods by mail.

Ward went to Chicago and with his small savings opened the nation's first mail-order company. He called it Montgomery Ward and Co. and distributed his first catalog—a single sheet listing his wares.

Four years later the sheet of paper had turned into a 150-page book and Ward was a multimillionaire.

Placing It

Some of the best ideas came from the strangest places. For example:

* William Hewlett and David Packard did their best thinking in a garage. They started tinkering with electronics in the shed behind Packard's tiny rented California home in 1938 and built a company, Hewlett-Packard, which in 1980 had sales of $3.1 billion and a work force of 57,000.

* Edson de Castro set up Data General Corp. on kitchen tables inside an empty beauty salon in Hudson, Massachusetts. The company sold more than $650 million worth of computers in 1981.

* A group of West Coast entrepreneurs met in a tavern over a few beers and came up with Genetech Inc. in 1976. It is now one of the leading companies in the development of new drugs and chemicals.

The Sun Also Rises

A bored soldier developed a chance idea during World War II that eventually made him a fortune.

Dan Lightfoot was stationed on Okinawa when he became fascinated with the idea of using the sun to heat things. His first solar heater was a network of garden hoses filled with water, but he kept refining his ideas.

More than 30 years later, with energy sources getting tighter and tighter, Lightfoot finally sold his ideas and made his fortune.

The Hole Truth

Chance can be cruel. Take the story of candy-maker Clarence Crane.

He was operating a small business in Cleveland, Ohio, in the early 1900s when his mint-stamping machine broke down. Crane fixed it, but noticed the mints wouldn't come out cleanly unless there was a hole in the middle.

Crane looked at the white mints and thought of a name—Life Savers. But he didn't know he was looking at a gold mine.

A young New York advertising executive named Edward John Noble saw a roll of Life Savers one day and bought the recipe and trademark from Crane for $2,900. Four years later Noble was a millionaire and Crane was still wondering what had happened.

The Man Who Passed Go

A man ruined in the stock market crash of 1929 turned his own misfortune into millions.

Charles Darrow was broke and out of work in 1933, so he started tinkering with an idea. Everyone was talking money and real estate but no one had either. Why not give them a game to play?

He worked on his idea and designed a game around money. He decided to call it Monopoly and a printer friend turned out the boards for him.

The little idea born in Germantown, Pennsylvania, grew to neighboring towns, and Darrow was pestered by people to turn out more games. He finally went to Parker Bros., who studied the game, at first rejecting it, and then on second thought refining it and bringing it out for Darrow.

Three years later Darrow was a millionaire. He was a multimillionaire when he died in 1970, and Parker Bros. has gone on to become the biggest game company in the world. They have turned out $1 trillion worth of monopoly money.

THE CASTLE **BUILT TO HONOR THE ACE OF CLUBS**
MidFord Castle, Bath, England.
ERECTED BY CAPT. HENRY ROEBUCK IN THE SHAPE OF THE PLAYING CARD THAT WON HIM THE FORTUNE IT COST!

Go West Young Man

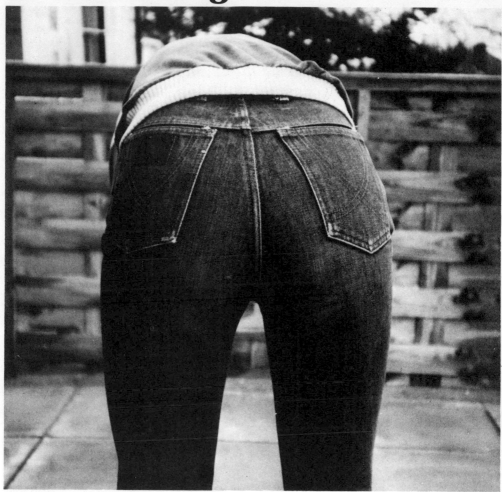

A penniless tailor who wanted to make tents took a chance instead and made millions.

Levi Strauss left his native New York in the 1850s to head west, where gold strikes were reported everywhere. He took with him most of his stock, but lost it on the way and arrived in San Francisco with only a roll of canvas.

He planned to make tents, but miners told him they'd rather have trousers because the rough woolens they wore wouldn't stand up to the strain of work. Strauss agreed and made one pair of trousers which became an instant hit.

Other miners came to Strauss and asked him for a pair. When they asked him what they were called, Strauss replied "Levis."

The name stuck, the trousers became a sensation and Strauss was working day and night. The miners wanted pockets that wouldn't rip, so Strauss decided to rivet them as a joke.

As his trade prospered, Strauss began looking for new sources of canvas. He found the perfect material in a factory in Nimes, France, and brought it back to America, shortening the name to "denims" ("from Nimes").

Strauss died a millionaire, but his Levis and denims will live forever.

A Fat Profit

A fat little housewife in Little Neck, New York, turned her diet plan into millions.

Jean Nidetch noticed all her friends had the same problem as she did, so she rounded them up to talk about their weight. That soon evolved into recipe exchanges, weekly announcements on weight loss and a general feeling of comradeship.

Why not share the idea, thought Nidetch, so she took a chance and incorporated a company called Weight Watchers International in 1963. The company has since enrolled more than 500,000 members around the world and is still growing.

Those Arches Are Golden

A curious salesman took a chance on efficiency and became one of the richest men in America.

Ray Kroc wondered why a tiny restaurant in San Bernardino, California, had placed an order for six milkshake-mixing machines when everyone else in the area only had one.

He visited the drive-in and discovered that not only was it selling more milkshakes, it also had 10 times the hamburger and soft-drink business. He watched for a couple of days and found that people liked its cleanliness and, most important, its speed.

Kroc also looked at its mass-production techniques and decided to buy the restaurant. Business stayed good and Kroc started selling franchises, which quickly spread across North America and into Britain.

Today, Kroc is one of the 10 richest men in America. And, oh yes, that little restaurant was called McDonald's.

Another Cuppa

Thomas Johnstone Lipton took chance by the neck and bent it his way.

In 1864, aged 12, the boy already showed his streak of genius when he shocked his father by saying that his mother should serve the eggs because they looked bigger beside her tiny hands.

The boy then journeyed to America, where the Civil War was raging, and learned one lesson: advertise.

He returned to Britain a brash young lad in 1871 and opened a shop in Glasgow, plastering his windows with sale stickers and organizing publicity stunts such as a parade of pigs. He even tried to give Queen Victoria a five-ton round of cheese.

They laughed at him, but they shopped at his stores and, at 29, Lipton was a millionaire. He then decided to import his own tea.

He traveled throughout Ceylon and India, buying up huge plantations and in 1895 made his mark when one of his ships ran aground in the Red Sea. Instead of bemoaning the fact, he painted "Drink Lipton's Tea" on the sodden bales and they floated throughout the Middle East.

The queen was not too sure about this crazy tea salesman, but a donation of £34,000 to her jubilee in 1887 bought him a knighthood in 1902. Later on, he became a baronet.

With Britons drinking tea nonstop, Lipton turned his attentions to yachting and spent the rest of his life trying to win the America's Cup for Britain.

Lucky M*A*S*H

A New York agent got a laugh from a book and decided to take a chance that made millions for him and Hollywood, established several big stars and launched one of the most popular television series in history.

Ingo Preminger read a comedy about the Korean War and telephoned Richard Zanuck, who had just taken over the sagging 20th Century-Fox film corporation. He listened politely, but paid no attention to Preminger.

The agent persisted, however, and Zanuck finally decided to make a movie out of the book, risking $3 million of the company's rapidly dwindling capital.

The movie was called M*A*S*H and returned more than 12 times its original investment. It established the careers of Donald Sutherland and Elliott Gould and was made into a television series.

And, oh yes, Preminger was producer and made millions of his own.

THE SAFETY PIN ONE OF THE MOST USEFUL INVENTIONS, WAS CREATED BY WALTER HUNT, A NEW YORK MECHANIC **BY ACCIDENT** — HE WAS IDLY TWISTING A WIRE WHILE TRYING TO THINK OF SOMETHING THAT WOULD ENABLE HIM TO PAY A DEBT OF $15

Give It Lots of Water

Gary Dahl dug up some rocks from his back-yard in Santa Cruz, California, and decided to take a chance on a joke.

A lot of people thought he was crazy, but Dahl put his rocks in a box, wrote a funny pamphlet, and became a millionaire. People bought Pet Rocks like crazy and Dahl is still laughing.

Blip Blip

Nolan Bushnell took a chance and $500 and started a little company in 1972.

He believed there was a future in electronic games that could be played on a television set. He patented a game called Pong and the next year his company sold $11 million worth of games.

A few years later Bushnell liquidated his assets in the San Francisco company for $28 million. But the company is still rolling along today.

Its name: Atari.

Quirks of Fate

Not all stories of invention have happy endings. Some inventors were never given credit for their work; others suffered great misunderstanding or misfortune. For example:

* Earl Laphrop, inventor of the world's first forgery-proof machine, was later sent to prison for life for forgery.

* King Henry II invented those huge frilled collars in France in the 16th century to hide a scar on his neck.

* Dr. Lee De Forest, the father of radio, was arrested and prosecuted for selling stock in his invention, which the judge called a "worthless piece of glass." His radio vacuum tubes made it possible for the first radio broadcast in 1910, a recording of Caruso.

* Everyone thinks Fulton invented the steam-boat, but it was actually John Fitch, who built a passenger vessel that operated be-tween Philadelphia and Trenton, New Jersey, in 1790. Fulton's boat didn't arrive until 1807.

* Shrapnel was invented by Gen. Henry Shrapnel, and he was the first man wounded by his own idea. It happened in a premature explosion at the 1793 evacuation of Dunkirk during the French revolutionary wars.

* Charles F. Dowd, the man who invented standard time and helped set up a network of railway time schedules in the U.S., was killed in a train accident.

Try and Try Again

Actor Kirk Douglas once bought an obscure little book about a mental institution but gave up trying to sell it as a movie.

Then his son, Michael Douglas, took over, badgering everyone in Hollywood until he finally got the capital from a rock musician. Douglas lured Jack Nicholson to be the star, and the film was made.

Incredibly, *One Flew over the Cuckoo's Nest* was a runaway hit and its cast and director swept the Academy Awards. Michael's movie made $100 million—more than his father's movies made in his whole career.

The 60-Second Click

One of the world's most famous cameras got on the market because a singer thought it was cute.

Dr. Edwin Land developed a photo process in the 1950s that instantly developed its own pictures. But no one wanted to take a chance on it because it looked too complicated.

Land finally got it manufactured, but sales were slow until one of his friends agreed to do a television commercial for him. That was Perry Como, and his low-key message captivated a nation.

Oddly enough, Land offered to pay Como in stock options, but he turned it down. He thought the camera was cute, but never believed people would actually buy it.

The camera was called Polaroid and Como has been regretting his stock decision ever since.

I'll Bet the Car

One reason Michelin owns Citroen today is Andre Citroen's love of chance.

The man was addicted to gambling, especially baccarat. His cars were popular throughout Europe, but at one sitting at the tables Citroen lost $400,000. He also lost the car company which still bears his name.

Ring, Ring

Chances are you thought Alexander Graham Bell invented the telephone.

Wrong. He not only didn't invent the phone, he was not even the first to design one. He was, however, the first to patent one.

In February 1876, Bell filed a patent for his telephone only a few hours *before* Elisha Gray filed an intent to patent a similar device. Thomas Edison was also working on his own phone when he became distracted by other work.

Today, the Bell patent is considered priceless — the most valuable patent ever filed in the world.

Foreign Affairs

Several people made millions because they took a chance in another country, lifting an idea that worked at home and setting it up elsewhere.

* Two young Americans went to Pretoria, South Africa, and built a hotdog stand. They made $25,000 in a month.
* Another American started a driving school in Tehran and made millions before the shah was deposed.
* A Nevada entrepreneur opened a root-beer stand in Malaysia, and on opening day, police had to control crowds. It grossed $200,000 in the first year.

How Not to Succeed

Everyone agreed the Edsel was the car of the future, the automobile that had every chance to succeed.

Named after Henry Ford's son, the model was the result of many years' work by the huge auto company. But it was brought out in 1957, when car sales across the country fell at the same time and the stock market took a sharp drop.

The car itself had problems with oil leaks, faulty buttons and shoddy workmanship. Ford finally withdrew it in 1959 and lost more than $350 million.

"No Edsel is dead," says Hugh Lesley of Oxford, Pa. Despite the huge loss Ford suffered on the car, Lesley defied the odds and collected 90 Edsels which now dot his 400-acre farm.

It's Fast, but Is It Profitable?

The Concorde may turn out to be the Edsel of the air.

Built to fly at supersonic speeds between Europe and North America, the needle-nosed plane was jointly financed by Britain and France. But from the start it had problems.

The jet was costly to fuel and arrived just as energy prices soared. It had many technical problems and its sonic boom was roundly attacked in North America as being a health hazard.

The two countries stopped building the planes in 1979 after only 16 had been completed, and now there's talk of abandoning existing flights.

Typewriter Trouble

Three men took the same chance in 1868 but only two turned out lucky.

The three invented the typewriter, and two of them made arrangements for royalties with manufacturers. But the third, C. Lathan Soles, decided to sell his idea to Remington for a flat $12,000.

NIAGARA FALLS WAS PURCHASED BY JEAN de COU FROM THE INDIANS IN 1788 FOR $5 AND A BLANKET. HE HOPED TO USE ITS WATER POWER TO OPERATE A MILL – BUT HIS INVESTMENT WAS A FAILURE

How to Make a Million

Your chances of making a fortune are guaranteed if you can come up with the following. Manufacturers agree they would love to be able to make:

* A small powerful storage battery that can run a car.

* More-convenient home movies.

* A pen that burns the words into paper

with no smudges, no ink stains and no possibility of erasure. Oh yes, it also has to be cheap.

* A new and better way to shave.

* A cheap way to reclaim silver from photographic film.

* A beer that isn't bitter.

* A plastic soda-container that doesn't explode.

* A small cheap helicopter.

* A steam iron that never needs filling.

* A seamless waterproof floor you can pour from a can.

Homemade Ideas

Look around the house: chances are you can discover a need for something that nobody has yet thought of. These people did and made a fortune in North America.

* B. F. Hamilton used a kitchen match to light his cigar and then didn't know where to put the burned match. He invented a handy little tin box with matches in one end and a receptacle in the other. That one idea helped form Hamilton Cosco Inc., the company that markets thousands of household items.

* A Baltimore druggist wanted to come up with something to "knock out eczema." He developed a salve and used those syllables to give us Noxzema.

* A Michigan housewife didn't think much of the cleaning products on the market so she mixed her own brew. It's still around and called Spic and Span.

* Leo Gerztenzang wanted to help his wife wash the baby so he came up with cotton swabs on sticks. He called them Q-Tips.

SETH POPE A PEDDLER OF Sandwich, Mass. WHO WAS ORDERED OUT OF TOWN BECAUSE IT WAS FEARED HE WOULD BECOME A PUBLIC CHARGE RETURNED **30** YEARS LATER *AND BOUGHT THE ENTIRE CITY*

Flying Saucers

The Frisbee was invented by a man who was bored with his job.

Fred Morrison noticed in 1947 that the newspapers were full of stories about flying saucers, so he decided to go to his garage and cash in on their popularity.

He developed a plastic disk with curled edges and called it a Pluto Platter, but the name was changed four years later when Morrison was approached by the Wham-O Manufacturing Corp. They settled on Frisbee, and in 1957 the first batch was produced in California.

Since then, Wham-O has turned out 100 million Frisbees. But it also had another big winner: the Hula-Hoop, originally designed by a health instructor in Australia as an exercise device.

Chances of Making a Million

Here are some interesting facts from the U.S. Internal Revenue Service:

∗ In 1974, out of 80 million tax returns in the U.S., only 903 people reported an income of over $1 million. By 1978, the number had risen to 2,041.

∗ Delaware has the most millionaires per capita of any state.

∗ Only 16 percent of a millionaire's income comes from salaries.

∗ Millionaires average $200,000 per year in donations to charity, all tax-deductible.

These are the hard facts. Now we look behind the scenes and take a peek at some of the incredible stories of people who took a chance and made that magic million.

SAMUEL COLT (1814-1862)
INVENTOR OF THE REVOLVER THAT BEARS HIS NAME, GOT THE IDEA FOR ITS REVOLVING CYLINDER AS A 16-YEAR-OLD SEAMAN WATCHING THE HELMSMAN TURN THE SHIP'S WHEEL--EACH SPOKE ALIGNING WITH A CLUTCH THAT HELD IT FAST

Call Me Madam

The Everleigh sisters, Minna and Ada, got lucky on the wrong side of the law.

When they married, their father gave them $40,000 each, but the marriages quickly turned sour and the two girls decided to become madams. They bought a house in Chicago and a stable of prostitutes, and decided their house would cater only to the very rich.

They filled it with a $15,000 gold piano, paintings and a vast library of classics and organized some of the nation's most sumptuous orgies. Word quickly spread and the Everleighs became famous and were visited by foreign dignitaries.

Strangely enough, both were staunch temperance enthusiasts and used to welcome Bible-thumpers who wanted to rescue their girls from a life of sin.

The Chicago police finally closed the house in 1910 and the two sisters ended their lives as retired millionairesses, living quietly in New York.

Among My Souvenirs

If you can't invent something, maybe you can make a fortune from that item you were about to throw out. The unlikeliest things have been big sellers in modern times and most were preserved only by chance.

* The garage wall that served as the backdrop to the St. Valentine's Day Massacre in Chicago was bought by George Patty of Vancouver when the building was being torn down.

* One of Winston Churchill's cigar stubs was sold at auction for $28.

* A frozen elephant carcass went for $1,500.

* A piece of Queen Victoria's wedding cake was sold for £68 by Christie's in London, despite the fact it was 133 years old.

* Pull-chain toilets, relics of the 1800s, go for $150 apiece in California.

* The first Superman comic was resold for $1,800.

* Original James Thurber cartoons go for around $500 each. Thurber loved to doodle these and then throw them in the wastebasket.

* P. T. Barnum tried—but failed—to buy Sarah Bernhardt's amputated leg in 1915.

* The town of Podunk Center, Iowa, is for sale at $7,000.

* Buffalo Bill sold an Indian scalp for $50.

* A New Jersey couple, James and Pamela Green, were arrested for trying to trade their 13-month-old son for a black and silver Corvette automobile.

Sarah Bernhardt (1845-1923).

Buffalo Bill staging a mock buffalo hunt in his Wild West show, 1887.

Winston Churchill, complete with cigar and tommy-gun.

...And for the Ladies

New York City has more millionaires than any other city on earth, and many of them are approachable. But two stand out as spectacular catches:

* Hugh Hefner, founder of *Playboy*, has around $25 million, but you'll have to compete with the Bunnies.
* David Rockefeller is still single at 40 and says he is trying to find out how to spend his life. He has $15 million to help him make up his mind.

The Inheritance That Can't Be Found

The Peruvian embassy contacted a California woman in 1949 to tell her she was a descendant of the Spanish viceroy who ruled Peru in the late 1700s.

The crusty viceroy, Antonio Pastor De Marin, had specified that his fortune not be divided until the fifth generation of descendants following his death.

Peruvian officials had a valid receipt of deposit for De Marin's fortune and it was all placed in the old Bank of Scotland. But the embassy is still trying to find the right bank.

The inheritance? More than £460 million or $1 billion.

Hello Out There

Chances are you've dreamed of the day when someone walks up to you, taps you on the shoulder and hands you $1 million just to do what you like doing.

Well, the $750-million MacArthur Foundation is doing just that.

Set up by John MacArthur, an eccentric billionaire who ran his empire from a Florida coffee shop, the foundation's job is to give away $37.5 million a year to free potential geniuses from humdrum jobs so they can think and work in peace.

And there are no strings. No reports have to be written, no applications have to be filed and the geniuses never have to say where the money went.

The Inheritance

One of the best ways to become a millionaire is to inherit the money, but you'd be surprised at the number of people who don't know they've been left it. U.S. bankers estimate $20 billion lies untouched in old accounts, stock certificates, safety deposit boxes and other legal depositaries, most of it destined to revert to state ownership if the rightful heir can't be found.

One Manhattan company, Missing Heirs International, claims to have found 80,000 people, delivering sums ranging from $1,000 to $7 million. Among the stories:

* A California man was told he was heir to $6 million and could collect it if he admitted he was an illegitimate child. The man refused.

* A 48-year-old prisoner at the Orange County Jail in California was told she was heir to a $4-million estate.

* Edwin Lewis Clark died in Veterans Hospital in Los Angeles in 1967 and the hospital assumed he was a pauper because he lived in a tiny apartment and had only two suits. Receipts left in a drawer, however, revealed that Clark had an investment portfolio that included $3 million worth of United Artists stock.

The Guggenheims

One of the richest families in the world is also one of the most eccentric. Peggy Guggenheim is best known for her fabulous art collection and huge donations, but she came from a mining family that, despite its millions, had some odd characters.

One of her aunts gambled compulsively and liked to sing loudly in public places. She had a phobia about germs and used to clean the furniture with lye.

The aunt's husband tried to kill her with a golf club. When he failed, he tied weights to his feet, jumped off a bridge and drowned.

Peggy's uncle liked to eat charcoal and ice cubes. He also drank heavily and ended up shooting himself.

Another uncle wanted to be a snake and used to slither around the house, scaring his children.

Peggy's father left his wife and lived like a gypsy in Paris. He died on the *Titanic* in 1912.

Peggy herself wanted a nose like a flower, but the plastic surgery of the time wasn't that good. Her newly tilted nose used to swell just before it rained.

Jackson the Loser

We've read many stories of starving artists, but none equals the curious relationship between Peggy Guggenheim and Jackson Pollock.

Peggy started by inheriting $900 million from her parents and discovered Pollock along the way. He was a struggling painter, eking out an existence as a carpenter and drinking all he could.

Peggy championed him, bought him a house and paid him $150 a month on which to live. But she couldn't peddle his work—some of which was done by throwing his whole body on the canvas and rubbing in the paint.

Pollock finally died, worn out by illness, and Peggy gave away 29 of his paintings, keeping only two. Then, chance struck, and the art world discovered Pollock, hailing him as the greatest painter since Picasso.

The 29 paintings Peggy had given away were worth millions and Pollock's widow, Lee, became a millionaire just by peddling the few she had left.

The Top 10

Chances are these could be the richest men in America.

* Ray Kroc, founder of McDonald's and owner of the San Diego Padres, is worth between $250 million and $500 million. If you rang his doorbell, the chimes would play "You Deserve a Break Today."

* Nelson Bunker Hunt likes to wear cheap chocolate-brown suits, although his wealth, which came mainly from oil wildcatting, was once estimated at $5 billion. Because of speculation, it has since declined, but Hunt can still find the odd $500 million or so.

* Edwin C. Whitehead started a little medical equipment company called Technicon and then merged his wholly-owned firm with Revlon for $500 million.

* Stephen Bechtel, founder of the world-famous construction company, has retired with his $400 million.

* Michael Fribourg is head of Continental Grain Co., which handles about a quarter of the world's shipping. His personal wealth is $400 million or more.

* Leonard Stern, owner of the Hartz Mountain Co., had made more than $500 million before he turned 35.

* Paul Mellon inherited his fortune from the banking family but now devotes himself to the arts and has given away more than $200

Daniel K. Ludwig, one of the richest men in the world.

million. His own fortune is estimated to be as high as $1 billion.

* Charles G. Koch was a Texas farmboy who invented a new way of refining oil. It worked, to the tune of $500 million.

* Forrest E. Mars, head of the candy company that makes Mars Bars, M and M's and Milky Way, is worth nearly $1 billion.

* Daniel K. Ludwig is recognized as the richest man in America. His fortune is between $2 billion and $3 billion and he has owned the largest shipping fleet in the world since he was 40.

What to Do with the Money

Chances are you've wondered what you'd do with a million if you had it. Here's what some people did:

* Marjorie Merriweather Post liked candy so much she kept a full-time candy-maker on her household staff.

* Charlotte Bergen periodically rented Carnegie Hall in New York and hired the American Symphony Orchestra to indulge her urge to conduct.

* Jefferson Seligman used to keep a closet full of fur coats to give away to cold visitors.

* Ned McLean, whose wife owned the Hope Diamond, hired a train from Washington to

New York and drove it himself, blowing the whistle.

* Listerine heir Gerard Lambert argued with his wife over whether their new house should face a lake or the ocean. Finally he built it on a peninsula so it had two views, constructed duplicate wings and visited her on the lake side occasionally to say hello.

* A British millionaire, angry at his wife's tantrums about his smoking, left her £500,000 when he died, provided she smoked five cigars a day.

* A Scotsman worth around £1 million left his wife only a handkerchief into which she could cry all day about the lost money.

When You Got It, Spend It

We've told you about millionaires and how they got there; now it's time for some tales about how some of them like to live it up. Chances are the rest of us will never match them:

* Michael Pearson, son of Lord Cowdray, the British head of a huge financial empire, inherited £10 million on his 21st birthday and once owned a large yacht called the *Hedonist*. He likes to wear cowboy boots and Stetson hats as he jet-sets around Europe.
* Keith Moon late member of the Who rock, group paid an estimated $500,000 just for wrecking hotel rooms after concerts.
* Singer Sammy Davis, Jr., has a $1-million home in Beverly Hills with 25 rooms and a film library of 1,000 movies. He has 25 television sets and five cars, one a $100,000 Rolls-Royce. He has a staff of 17 and they cost him $30,000 a week, but he likes to give lunches and pick up $3,000 tabs.
* Sheik Taufiq Aziz of Qatar bought three Aston-Martin Lagondas at the London auto show in 1976. The price? More than £60,000 apiece.
* The governor of Riyadh lost £3 million in three nights while gambling at Ladbroke's in Britain. When he walked out he tipped every waiter £150.
* Crown Prince Fahd of Saudi Arabia, brother of King Khaled, has his own jumbo jet. He once amused himself in September 1974 by trying to break the bank at Monte Carlo, but gave up after losing nearly $17 million.
* The sultan of Oman had a little birthday party for himself in 1976 and flew in the Grenadier Guards, four Indian elephants, Gerry Cottle's circus, a 15-ton tent, a 2,500-seat grandstand, five lions, two tons of straw and all the acts, including a Persian strongman and a motor-cycle high-wire act.
* Adnan Mohamed Khashoggi, the head of the multinational Triad Holding Corp., travels the world in his own Boeing 727, which has a 40-foot sitting room, two wardrobes (one of suits, one of gold robes), gold-plated toilet seats and a satellite radio hookup to allow him to talk to any place in the world. He also has three yachts, including a 350-footer, and 12 palaces.

Marrying Money

There are other ways of striking it rich. You could always marry money if you don't want to dig, dive or prospect. All it takes is luck and the right bride. Here are some of the world's most eligible women:

* Josephine Abercrombie is worth between $200 million and $300 million thanks to her inheritance—the Cameron Iron Works, which manufactures oil-drilling equipment.
* Muffie Bancroft Armory has around $5 million and lives in New York where she works as a model. She has never married. She inherited her money from her mother, who was part of the Standard Oil family.
* Anne Ford, daughter of Henry Ford II, lives in New York with her $25 million and is divorced from an Italian stockbroker.
* Gloria Vanderbilt, the great-great-granddaughter of Commodore Vanderbilt, has had four husbands and is worth up to $10 million.
* Katharine Graham, owner of the *Washington Post*, is conservatively estimated to be worth nearly $150 million.
* Cordelia Scaife May Duggan, a descendant of the Mellon family, lives in Pittsburgh with $200 million.
* Jacqueline Kennedy Onassis, widow of the former U.S. president and Aristotle Onassis, has accumulated an estimated $30 million over her 52 years.
* Happy Rockefeller, in spite of a stormy marriage to Nelson, has 50 million reasons to be happy.
* Ruth Jane Hunt, daughter of billionaire H. L. Hunt, is worth $50 million and is a born-again Christian who organizes gospel tours.
* Jane Engelhard, the widow of Charles Engelhard, lives in New Jersey with a fortune estimated at $300 million.

The Millionaire's Club

Want to make a million? Talk to Joe Cossman. He made his $1 million by listening.

After World War II, Cossman returned home to the United States with $276 and an idea. He parlayed that into a mail-order company to supply items badly needed in Europe. It quickly became one of the largest in the world.

He then wrote two books, *How I Made $1 Million in Mail Order* and *How to Get $50,000 Worth of Services Free Each Year from the U.S. Government.* That started an average 20,000 letters a year from people, so Joe decided to set up a chain of franchised Future Millionaire clubs.

These seek out individuals with marketable products and show them how to test, develop, produce, sell, advertise and promote their products. His first club in Los Angeles quickly signed up 800 members.

Cossman's own philosophy: Never tie yourself to a weekly paycheck. "Give a man a fish and you feed him for a day; teach a man to fish and you feed him for life."

There's No Place like Home

Chances are you'd be richer if you lived somewhere else. The United Nations has prepared a per capita list of the world's wealthiest countries:
1. United Arab Emirates
2. Qatar
3. Kuwait
4. Liechtenstein
5. Switzerland
6. Sweden
7. Monaco
8. United States
9. Canada
10. West Germany
11. Australia
12. Denmark
13. New Caledonia
14. Luxembourg
15. Belgium

But Is It Art?

Inflation and devaluation of the dollar have made chances of getting rich in the art market much better. Here are some spectacular examples of people who gambled and won:
* In 1950 a Rembrandt portrait of the artist's son Titus sold for $160,000. But in 1965 it was sold again for $2.2 million and today is worth more than double that.
* Etchings by Goya went for $5,000 in the 1950s, but the price 10 years later was $58,000. Today, the sky's the limit.
* Jackson Pollock abstracts seldom went for more than $1,000 in the 1950s, but 10 years later they were selling for $100,000 and more.
* Andy Warhol sold a painting in 1962 for $200, but the canvas—a collage of 200 dollar bills—is worth $20,000 today.

If You Got It, Keep It

A millionairess lived like a pauper.

Hetty Green, born in Massachusetts in 1835, inherited a substantial fortune when her father died and she hung on to almost every penny.

When her son broke his leg, she did not call a doctor and instead took him to a charity hospital where she passed herself off as a beggar. The boy's leg worsened and had to be amputated, but Hetty had it done in her rooming house to save hospital bed fees.

In later years, she lived in an unheated tenement and subsisted on a diet of cold eggs and onions. She wore newspapers for underpants and only washed the lower half of her dresses.

When she died in 1916, her estate was probated at $125 million.

Insurance

Millionaires and large companies have quickly discovered that the element of chance can be cut down with insurance. Here are some stories of claims and policies.

* The entire set for a shot in the movie *Superman* was ruined by a power blackout, and the insurance companies shelled out $50,000.
* Another set painstakingly built for the movie *Apocalypse Now* was wiped out by a typhoon, and the film company collected $1.5 million.
* The television series *Chico and the Man* collected $125,000 a week from insurance companies after star Freddie Prinze committed suicide.
* United Artists received $2 million when Peter Sellers died during preproduction of a new movie.
* The average Broadway show carries $750,000 to $1 million in insurance against not opening; musicals carry $2 million.
* Elizabeth Taylor was insured in *The Little Foxes* and was ill for 11 days. The producers collected $110,000.
* Edy Williams, star of soft-core porn films, insured her breasts.
* Betty Grable, on the other hand, insured her famous legs.
* Ben Turpin, the famous silent-film comedian, insured his crossed eyes.
* Douglas Fairbanks, swashbuckling star of the silents, protected his face against scars because he did most of his own stunt work.

* Sabrina, a well-endowed British actress of the 1960s, insured her 41-inch bust against everything but war, nationalization and invasion.
* Robert De Niro put on 60 pounds for the film *Raging Bull*, but took out insurance for his health.

Sir John Mansell, King's Counsellor, Royal Secretary, and Chief Justice of England, held 300 salaried government posts at the same time. He was known as the richest clerk in the world, and his influence eventually so enraged his countrymen that they finally forced King Henry III to dismiss him. He died in poverty in France in 1265. Believe It or Not!

Keeping up with the Joneses

Your chances of going bankrupt are much higher if you live near a person who has already gone into receivership. Almost 28 percent of the bankrupts in one American city lived on the same block and another 28 percent lived two blocks away.

With this in mind, credit companies and banks take a long hard look at the people they lend money to and have found that:
* Business executives pay their bills most quickly, farm laborers most slowly.
* School teachers and clergymen are poor credit risks. Policemen and firemen are even worse.
* Lawyers and judges are also slow to repay debts, but bartenders and painters are considered the chanciest of all.
* Accountants, store managers, doctors, engineers, clerks, college professors and railway clerks are considered the best risks.

Pay or Die

If you had lived in ancient Rome, the odds are you wouldn't have taken bankruptcy lightly. Back then they took the creditor's side.

Two popular ways of punishing someone who couldn't pay his bills were to pull him apart with a wild horse or cut him up with a dull ax. In the latter case, the biggest creditor got the first whack.

King Henry VIII had a marvelous debt collector named John Dun. He was so good at his work that the term "dun him for it" is still used today.

Debtor's prisons, by the way, were not abolished in Britain until 1896.

L.L. ROCKWELL OF Fort Recovery, Ohio
IN 1902 PUT HIS INITIALS
ON A SILVER DOLLAR AND SPENT IT —
IT CAME BACK TO HIM IN JANUARY 1937-
AFTER BEING IN CIRCULATION **35** YEARS

The $5-Million Eggs

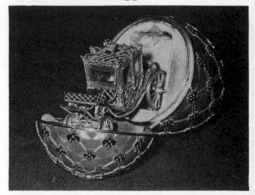

Chances are you'd never think an Easter egg was worth a fortune. But it can be—if it was designed by Faberge, court jeweler to the czar of Russia. He presented a royal egg to court each Easter, decorating the shells with pearls, diamonds and rubies. Inside were miniature crowns and rings, picture frames and platinum swans. One egg even held a small train, perfect in every detail. He made 57 eggs in total and 53 exist today, worth an estimated $5 million.

The Credit Card Jungle

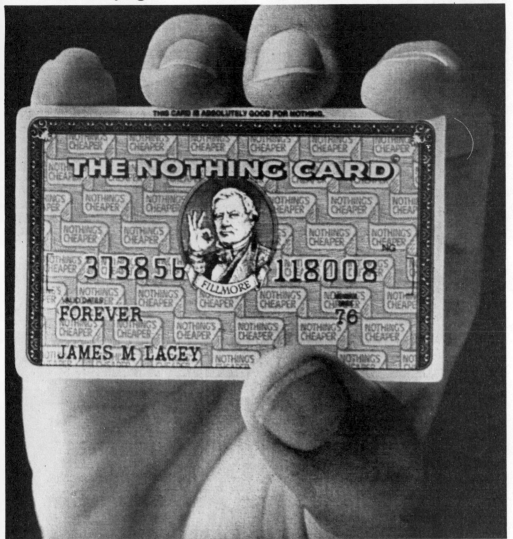

If you live in the Western world, chances are you're in debt to one or more credit card companies.

Mastercard and Visa handle 75 million accounts around the world and do $25 billion worth of credit business each year. In the last few years, Americans owed some $223 billion to banks, retailers, credit card chains and credit unions.

There are more than 300 million credit cards in circulation and the average person in the U.S. carries four cards in his wallet.

How Credit Is Given

Generously. Your chances of getting a credit card—whether you want it or not—are good if:
* You've had your present job for 10 years.
* You've lived at your present address for two years or more.
* You've got a telephone in your name.
* You earn more than $12,000 a year and have a bank account.
* You are between 20 and 30, or 45 and older.

Going for Broke

Short of money, wondering how to make ends meet? You're not alone. Even the Vatican predicts a $25.6-million deficit for 1982.

Chances are you're living beyond your means. But even worse off are those who end up bankrupt.

Making Ends Meet

Chances are overwhelmingly against new businesses. In any given year, one in four will fail; over two years, one in two; over 10 years it's four in five.

But only 20 to 25 of every 100 new restaurants fail. And 99 out of 100 jewelry stores will still be open a year after they start business.

Junk Mail

It floods through the mailbox every day. The chances are 26 in 100 that any given piece of mail in the United States is an advertisement.

Does it work? Chances are one in 50 that a person will order a product advertised by mail.

Oh, Oh

Chances are you're on the way to bankruptcy if:

* You spend 20 percent or more of your take-home pay on loans and credit card payments.
* You're looking for a consolidation loan to pay off your debts.
* You pay one creditor one month and another the next month because you can't manage all at once.
* You're starting to get second-notice cards from creditors about overdue bills.
* You have to borrow to pay household expenses because all your cash is gone.
* You pay the minimum on credit card balances.

Bankruptcies: A Final Word

If you're going bankrupt, you're not alone. In America in 1980, 355,000 businesses filed as failures.

March, incidentally, is the worst month for bankruptcies, and creditors in such cases get back only 10 cents on the dollar.

Is there a bright side? Maybe. If you're bankrupt in the U.S., you get a second chance and are allowed to keep $7,500 of your personal assets.

The Collectors

Your chances of having a loan repaid are excellent if you're a bank or mortgage company. They're good if you're a finance company, a credit union or a retail store.

They're fair to good if you're a doctor, a dentist, a pharmacist or a hospital. But if you own a gas station, payments can be very slow.

And if you've lent money for a specific purchase, chances that it will be repaid are excellent if the loan was for a mortgage. People also repay money borrowed for new cars and television sets. But if you lent money for clothes, don't hold your breath.

Portrait of a Bankrupt

Chances are the average bankrupt fits this profile: He's male, Caucasian, aged around 30, married with three children, has little education and at least 18 creditors.

An Oregon study showed that out of 100 bankrupts, 14 people had deserted their wives or families, 10 had passed bad checks, 4 had committed larceny, 2 had committed burglary, 2 were arrested for fraud and 1 had committed rape.

ROBERT MORRIS

WHO SUSTAINED THE FLEDGLING U.S. GOVERNMENT WITH HIS PERSONAL FUNDS, WAS BANKRUPT IN HIS FINAL YEARS, SPENT 3 YEARS IN A DEBTORS' PRISON *AND DIED A PAUPER*

The White House

Looking for the biggest job in America—the presidency? Chances are excellent you'll make it to the White House if:

* You're a lawyer. Twenty-four out of 39 presidents were.

* You attended college. Only 10 didn't.

* You have a name ending in "son." Eight presidents did.

* You went to Harvard. Five did.

* You're not a draft dodger.

Plastic Charge

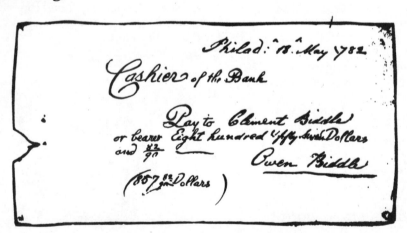

Despite computerization, the chances are good your credit card company will make an error. Consider:

* A Florida man used his credit card as a reference and was turned down because the computer showed legal charges against someone with the same name.

* Anna Lindstrom got 12 dunning letters asking for repayment of a balance of $0.00.

Finally, in desperation, she sent off a check for $0.00. She got a thank-you letter.

* A dachshund named Alice Griffin got four credit cards after doing some work in dog-food ads.

* Josephine DeVoto of Cambridge, Massachusetts, received two credit cards for her dead husband.

Get a Job

Jobs mean money, right? And millions of people each year take a chance on a new career, a new future or a change in work. But the element of chance is more important than qualifications in a lot of cases.

* One out of every four U.S. college students winds up in a job that doesn't require a degree.

* If the civil service in the U.S. continues to grow at its present rate everyone will be working for the state, local or federal government by 2049.

* About one in 100 job applicants lies about having a college degree.

* Only half the teachers who have graduated since 1979 have found jobs.

* There are two jobs waiting for every petroleum engineering graduate in the U.S., and the starting salary, regardless of age, is an average $22,000 a year.

Yes, but Can You Type?

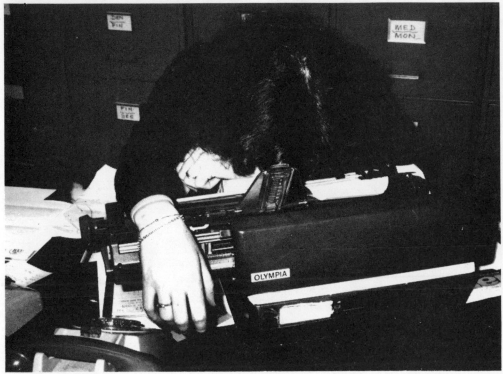

Qualifications are an important part of getting a job, but chance has laughed at the following:

* The man who posed for the "I Want You" poster for the U.S. Army was declared ineligible for a pension.
* A weight lifter who became Mr. Seattle because of his marvelous body couldn't pass the physical to become a policeman in the same city.
* Desperate for officers and lacking applicants, the Miami police department lowered eligibility requirements. If you want the job, you can get it if you can prove you haven't smoked marijuana more than 15 times and haven't used amphetamines or barbiturates more than 10 times. Heroin and cocaine users are still out of luck.

All in a Day's Work

We turn now to odd facts:
* Statistics show that if a business executive hears a rumor, chances are three in four it will be true.
* On any given day in North America, if an employee is absent from a company, chances are 56 in 100 he or she is a smoker.
* Nine out of 10 company thefts are done or abetted by employees. And stats show that retail clerks steal about nine times as much as shoplifters.
* If a piece of paper is filed, chances are one in 20 that it will never be taken out again.
* Out of every 100 paper clips, 25 will go in the garbage, 20 will clip something together, 19 will be used as poker chips, 14 will be bent out of shape, 7 will be used to hold up a seam, 5 will be used to clean nails, 5 will be toothpicks, 4 will clean a pipe.

Dangerous Jobs

Thinking of changing jobs? Chance would be against you if you decided to become an astronaut, the single most dangerous occupation in the world.

Right behind that vocation are race-car drivers, boxers, electric workers and steeplejacks. And if you live in America your chances of injury are good if:

* You're a teacher. One in 20 is attacked by students each year.
* You're a policeman. One in 5,000 gets killed yearly.
* You're an income tax inspector. Chances are two in five you'll be threatened.
* You're a mailman. Chances are one in 16 you'll be bitten by a dog.
* You're a unionized actor. Chances are only one in four you'll make more than $2,000 in a year.

If those don't shake you, how about job suicide statistics? On average 4 people in 10,000 kill themselves yearly, but rates are higher among pharmacists (12 in 10,000), dentists (8.3), doctors (7.9), lawyers (5.4) and engineers (4.5).

Where the Jobs Are

Companies, universities, high schools and governments all agree your best chances for work in the future will be in:

* Information technology, or how to love a computer.
* Biotechnology, including chemistry, genetic engineering, agriculture and agronomy.
* Small businesses and self-employment companies that specialize in repairing computers.
* Franchising, especially in hotels.
* Law, medicine, teaching and professional soldiering.

NEW YORK CITY HAS SOME 7,000,000 BATS YET HAS A BAT PATROL CONSISTING OF ONE WOMAN

Big Brother

With every pot at the end of the rainbow, there's a tax man waiting to take his share of the treasure. Some more facts from an organization that doesn't believe in chance—the U.S. Internal Revenue Service.

* If you make $250 million or more, your chances are 100 percent of being audited.
* If you're only making $100 million, your chances are three in four the government will take your return apart.
* If you make less than $100,000, however, you have only four chances in 100 of being examined.
* If you make less than $50,000, chances are excellent the government will accept your return as filed.
* The U.S. government lost $26 billion in tax money in 1979 because of $136 billion in unreported income—people who took a chance and beat the system
* Colonel Sanders was a believer in the system, so he tried to deduct his white suits as a business expense for his Kentucky Fried Chicken enterprise. The government turned him down.

Bulls and Bears

The market is a roller coaster of chance around the world. Fortunes have been made and lost; penny stocks have turned into $100 investments; people, banks and governments have been ruined. But still the market is out there, ticking away, the barometer to the financial health of any country, company, individual or idea.

But with the euphoric highs there are the lows. And the North American stock markets have taken two serious beatings. The first came in 1929, heralding the Great Depression; the second came in 1969, introducing the first major recession. The latter was a sign of things to come—a world choked with inflation.

Both falls had parallels and we can all learn a lesson from them:
* In each case, the crash was said to be impossible. In 1929 the Federal Reserve System in the U.S. was supposed to hold up the market, and in 1969 the securities commission was supposed to do the same.
* Both crashes happened because of widespread and senseless speculation for paper profits. In 1929 it was margin players, and in 1969 it was mutual-fund operators trying to protect small investors.
* In both crashes, a Republican president had just been elected the year before. It was Hoover in 1928 and Nixon in 1968, both known as fierce protectors of private businesses.
* In both, there was an enormous buying panic to get in on a "good deal," followed by panic selling when inflated stocks suddenly collapsed.

Psst, Here's a Good One

Frederick Goldsmith ran one of the most popular tip sheets to the stock market for 40 years and many of his best clients were brokers.

But Goldsmith told the U.S. Securities and Exchange Commission in 1946 that the real source of his "inside" information to Wall Street came from the Maggie and Jiggs comic strip.

If Jiggs had had a bad day at work, Goldsmith would tell his thousands of clients to pull back in the market. Under later questioning, Goldsmith also said the ghost of financier J. P. Morgan visited him at night and whispered tips into his ear.

Easy Come, Easy Go

Want to play the market? You need more than chance on your side.

H. Ross Perot made $1 billion on the market before he turned 38, when he went public with a company called Electronic Systems. He owned the company and offered only 350,000 shares at $16.50 each.

But this was the mid 1960s when everyone wanted to jump on the computer bandwagon and Perot's company seemed better than most. The shares shot up to $130 in a year and Perot, who had kept nine million for himself, was a billionaire.

But by April 1970 the recession sell was on, and Perot lost more than $445 million in a year. In one day, he lost more than the entire welfare budget for any city in America except New York.

An Offer He Couldn't Refuse

Charles W. Bludhorn thought he had it all in the 1960s, but only chance saved him from financial disaster.

A major player on the stock market during the boom years of the '60s, Bludhorn bought Paramount studios. The market crash then hit and his paper holdings dropped from an average of $64.50 to $9. Almost ruined, he turned to his movie-making venture.

He liquidated the real estate holdings of the company, but was still against the wall when producer Robert Evans walked in with the script to *Love Story*, which made $80 million.

If that wasn't enough, Bludhorn reluctantly committed $6 million to make *The Godfather*, which went on to make $120 million and save the company.

Lucky Jim

Diamond Jim Brady, the colorful American financier, believed he was about to die in 1912 so he burned $20,000 worth of promissory notes owed him by his friends—and then recovered his health!

The Russian gold reserve in the State Bank in St. Petersburg, 1905.

If You Don't Like It, Move

Ever wonder what are the most expensive places in the world in which to live? The cheapest? Chances are there are a few surprises on these lists, compiled by the United Nations:

Least Expensive
Moscow
Colombo, Sri Lanka
Belize
Alexandria, Egypt
Valletta, Malta
Cook Islands
Bratislava, Czechoslovakia
Tarawa, Gilbert Islands
Kingston, Jamaica
Mexico City

Most Expensive
Tokyo
Geneva
Kinshasa, Zaire
Brussels
Manama, Bahrain
Bonn, Germany
The Hague, Netherlands
Copenhagen
Muscat, Oman
Djibouti

The Sweet Stuff

Chance fortunes have been made and lost on the humblest and most ancient food of all—sugar.

The Greeks and the Chinese first started speculating in sugar and then it was gradually replaced by honey. The Crusaders rediscovered sugar in Tripoli in the Middle Ages and since then, some bags have been worth their weight in gold.

Queen Elizabeth I was so proud of her royal warehouse full of sugar that she listed it as a major asset in her will. But it was not until 1887 that the juggling of sugar prices really came into vogue.

An American trader named Henry Havemeyer consolidated 17 companies and formed the American Sugar Refining Co., giving himself control over a nation's breakfasts. He used to raise the price at will and once said, "Who worries for a quarter of a cent a pound?"

Fortunes were made in American Sugar stock during the early 1900s when the House of Representatives debated whether to raise the tariff on sugar.

Then there's the story of Julio Lobo, the greatest sugar trader of modern times. He made his fortune in pre-Castro Cuba and built it to $200 million at one time. For 15 years after World War II, he was available to buy or sell any amount of sugar anywhere in the world.

He eventually cornered the market and before U.S. buyers could blink, he had driven the price up from $5.25 to $7.58. In one week, Pepsi-Cola bought carloads at between $5.20 and eventually $5.60 a pound.

But the bubble burst when Castro came to power. The Cuban dictator quickly confiscated Lobo's holdings, and in 1964 the ax finally fell. Lobo lost $5.6 million in one week when he gambled that prices would go up.

I Coulda Been Somebody

Do boxers prosper after they give up the ring? Chances are they don't.

Researchers studied 95 professional fighters who had earned more than $100,000 in their career and who had either held a championship or been a championship contender.

In the 95, they found 28 bouncers, tavern owners or bartenders, 18 laborers, 2 wrestlers, 2 janitors, 2 bookmakers, 3 taxi drivers, 19 house painters, 3 newsstand operators, 2 liquor salesmen, 2 gas station attendants, 1 tailor, 3 race track hotwalkers, 1 business-man—and the rest were fight trainers.

Patent Truth

So you want to be an investor? Chances are that patent is not as valuable as you think.

First, the average time to process a patent was 22.8 months in 1980 and 19.4 months in 1979. Of all the patent documents on file, at least 7 percent are missing.

The value of patents, particularly for a small-time inventor, is questionable. More than half those tested in court in 1979 were declared invalid.

The $2 Curse

The U.S. Treasury still prints them, but the American $2 bill is considered bad luck by millions.

The whole idea, it is believed, was started by gamblers who called a $2 bill a deuce, which became a slang word for Devil and bad luck. To take away the curse, gamblers used to tear off a corner because it formed a triangle, which was good luck.

It's not uncommon to find a $2 bill with all four corners torn off, but again another superstition rises. If you've got the cornerless bill, you're supposed to tear it all up.

Cashiers today still ask customers if they mind taking a $2 bill and some even kiss it for luck before handing it away.

Odd Tales

Chances are you didn't know these money stories:

* Florida State University ran a test using a new gold Cadillac and an old Ford. They left on the lights of both cars and passersby turned off the Ford's lights 82 percent of the time and the Cadillac's only 39 percent.

* The First Security Bank in Boise, Idaho, accidentally put 8,000 checks worth $840,000 through a shredding machine. It then hired 50 temporary employees to work six hours a day each to paste the checks back together.

* Peter Minuit, who bought Manhattan Island for $24, was fired by the Dutch for his extravagance.

* Horatio Alger made a fortune writing 119 books inspiring poor boys to labor diligently and save their pennies. He died in poverty because he himself became a spendthrift.

PART FOUR:

Blind Forces

Hazards, Marvels of Invention and the Natural World

Mother Nature has many faces. Sometimes she is kind and benevolent, showering blessings upon mankind. But more often she appears to be cruel, vicious and utterly unpredictable—playing with us like "flies to wanton boys," in the words of Shakespeare.

One of the greatest wonders of the world is man himself and the planet he lives on. That everything should have combined to produce exactly the right conditions for life to start is truly miraculous!

Where did we come from? What is going to happen to us in the years ahead? Are we alone in the universe? These are just some of the fascinating questions we tackle in this chapter— and the answers may surprise you!

Here are the great disasters of nature—earthquakes, fires, hurricanes, avalanches. Here are incredible tales of escape and man's unquenchable will to survive! Here are man's greatest achievements—stumbled upon in many cases by sheer chance.

Here then are the many moods of Nature—benign and gentle, and red in tooth and claw!

How on Earth...?

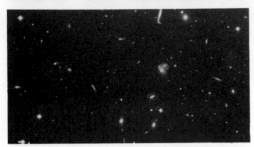

Should we be here at all? The odds were very much against it right at the beginning. Who would have bet that circumstances would combine in just the right way so that the earth itself would be formed, so that life would develop, so that man would evolve into the dominant species? The chances of it all happening are indeed astronomical—unless you believe in a divine being who caused it all:

* Having swept up most of the material in its orbit, the earth began to stabilize. A whirling cloud of gas and dust slowly evolved into our sun, which began to heat the mass of the earth, which by chance was covered in water.
* In the lifeless ocean, certain chemical changes began to take place, creating a "primordial soup" out of which life was created. Before the earth was a billion years old, it was a living world!

When the Sun Goes Out

"The laws of physics have written our eviction notice," says Wallace Tucker of the Harvard-Smithsonian Center for Astrophysics. "Before the dying sun becomes a red giant five billion years from now, we must vacate our planet."

We can hang around Earth for another three billion years, but after that we'll have to make our way to other planets as the sun expands and grows brighter. Then the sun will begin to collapse and we'll have to move on again!

By which time, of course, we may have found new ways to travel into outer space and settle down there. We may even have chanced upon a way to move Earth itself to somewhere safer!

In the Beginning

The universe is expanding. Distant galaxies are receding, at the same rate in all directions. And scientists have come up with two main theories to explain how it all began:

* The Big Bang. The universe began at high density, exploded, and scattered matter far and wide.
* Continuous creation. A few scientists hold to the idea that the universe is of infinite age and that matter is created at a rate that balances the expansion of the universe.

The Water of Life

The first thing a Martian would notice about Earth is not earth but water. This miraculous combination of elements is the basis for all terrestrial life.

* All nutrients for plants and animals must be dissolved in water to be digested.
* Our breathing and digestive systems and blood are largely made up of water.
* It covers 79 percent of our planet yet is rare on other planets.
* As vapor, it shields Earth from lethal cosmic rays.
* Life would never have begun on Earth without water and would quickly die if there was no water.

The Future: A Bleak Prospect

Man is lessening his own chances of survival on this planet. Even ruling out a nuclear war, which would end life on earth as we know it, there are numerous other ways in which man is stacking the odds against himself.

The Greenhouse Earth

With our increasing demand for energy, we are burning more and more fossil fuels. This in turn increases the concentration of carbon dioxide in the atmosphere, increasing the intensity of the sun's rays. Even an increase of five to 10 degrees in the earth's temperature would be catastrophic; glaciers would melt and the world's oceans would rise by about 300 feet, swamping most of the earth.

Signor Perni demonstrating his planetarium in London, 1880.

Flights of Fancy?

Reports of Unidentified Flying Objects have been around almost as long as man himself. An Egyptian writing 3,500 years ago described a mysterious flying circle, high in the sky. There are similar reports from ancient China, as well as from Roman and medieval times.

But today, the odds are that your mysterious saucer-shaped object can be explained — scientifically. Of all sightings, only 1 or 2 percent remain a genuine puzzle after the experts have got to work.

Some common "UFOs":

* Space satellites, or pieces of them, returning to Earth.
* The planets Mars, Saturn, Jupiter and particularly Venus.
* Weather balloons, which change shape as they rise.
* High-flying aircraft.
* Meteors, ball lightning and clouds.
* Many birds and insects.

The statistics are against UFOs. Investigator Philip Klass has pointed out that the nearest star system to ours that could have life on it is Alpha Centauri. It would take them nearly 100 years to make a round trip to Earth — supposing that their spaceship could travel at 70 million miles an hour!

Klass has laid several bets of $10,000 on UFOs. The *National Enquirer* has offered $1 million to anyone who can prove that UFOs come from outer space. Neither has had to pay up yet.

Calling All Planets

Carl Sagan figures we should make contact with aliens within our own lifetime. If we do, it will be sheer luck—our electronic "ears" will be tuned to the right place at exactly the right time.

In fact we've been tuning in to the heavens since 1960. And we've been sending out our own messages too. The *Pioneer 10* and *11* spacecraft each carried plaques with etchings of human images and a host of scientific data past Jupiter and Saturn and on into space. And the *Voyager 1* and *2* spacecraft each carried gold-plated copper records with samples of Earth languages from English to Esperanto, as well as music, the sounds of nature, calls of hyenas, whales and human babies, and the beat of a human heart. They also carried a number of coded pictures of DNA, the human anatomy and much more scientific information.

The world's largest radio telescope at Jodrell Bank in England.

The UFO Flaps

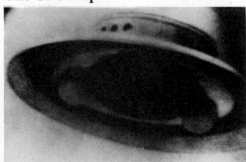

The first rash of UFO sightings started in the U.S. in 1896—just about the time inventors were pushing ahead with plans for the first flying machines. There were hundreds of reports:

* One California man claimed he heard the crew of a "mysterious airship" singing "Nearer, My God, to Thee."

* A Michigan man claimed he was awakened in April 1897 by the hungry crew of an airship, asking for a corkscrew, four dozen egg sandwiches and a pot of coffee. He gave them everything except the corkscrew.

The flood of reports stopped suddenly at the end of 1897.

Is There Anybody Out There?

The odds are that we are not alone, according to astrophysicists Carl Sagan and I. S. Shlovskii in their book *Intelligent Life in the Universe*. There are billions and billions of stars and the mathematical probability is that there are a million planets in our own galaxy, the Milky Way, that support intelligent life.

Others, such as the eminent British astronomer Sir Bernard Lovell, disagree. So many factors had to combine to produce the right environment for life on Earth that Lovell concludes "the chances of the existence of man on Earth today, or of intelligent life anywhere in the universe, are vanishingly small."

Closing the Blue Book

The U.S. Air Force started an investigation of UFOs in what came to be known as Project Blue Book. By 1969, 12,618 sightings were recorded as having been identified while 701 were marked "unidentified." The project ended that year when the air force announced that all the sightings could be explained. The British Ministry of Defence, which ran its own investigations, reached a similar conclusion.

But, there's still a chance...

Oh Lucky Man

Early man was an unlikely creature to inherit the earth. Originally a strain of climbing ape, evolutionists say, he left the trees and began to live on the grasslands, changing from a weaponless vegetarian into a clothed, mobile hunter who was to become the most feared animal on earth. There were several key factors:

∗ He had an opposable thumb, which enabled him to make weapons.

∗ He began to walk upright, which improved vision and hand-eye coordination.

∗ He discovered how to use fire and lived in caves for shelter.

Big, Bigger, Biggest

"Dinosaur Jim" Jenson, of Brigham Young University in Utah, is the most successful discoverer of dinosaurs in North America. In 1972 he unearthed an eight-foot shoulder blade and 10-foot ribs belonging to a creature he dubbed Supersaurus—70 feet long and weighing 70 tons. Then in 1979 he discovered Ultrasaurus. With a shoulder blade nine feet long, Ultrasaurus is 25 percent larger than any dinosaur on display in any museum!

Heads Will Roll

Giant brontosaurs have worn the wrong heads in museums for 100 years! And it's all because paleontologist Othniel Marsh, who discovered the first brontosaur in the 1870s, failed to mention that its snub-nosed skull was found miles from the rest of the skeleton. Only recently have dinosaur experts confirmed that the brontosaur had a longer, flatter skull!

Dinosaur Graveyard

The largest concentration of dinosaur fossils in the world is in Dinosaur Provincial Park in Alberta, Canada. There, 70 to 85 million years ago, dead dinosaurs floated downstream and became trapped and buried in mud.

Digging Dinosaurs

Dinosaur hunting is booming. Nearly 300 types have been named since the first dinosaur fossils were discovered in the 19th century,

The Monster Mystery

Sixty-four million years ago, dinosaurs vanished abruptly from the face of the earth. Was it due to natural evolution, the rise of mammals or a gradual climatic change? Or was it a chance disaster?

One of the newest theories is that a large meteorite or asteroid cannoned into the earth, raising a dust cloud that blocked out all the sunlight. The temperature dropped and when the dust had settled, all the dinosaurs had disappeared, along with many kinds of plants, plankton and mammals.

Who Rules the Waves?

Pleisiosaurus and Ichthyosaurus.

Man may think he's the only intelligent species on land, but things might have been different if fate had decreed it. Dolphins and whales, both highly intelligent mammals, once were land animals with legs. But for some unknown reason they returned to the oceans millions of years ago!

A Chance Hit

Halley's comet in 1910.

Earth's atmosphere is bombarded by 75,000,000 meteors every 24 hours! Mercifully, most burn up before they can land. Perhaps 10 a day actually make it to earth—but since three quarters of the planet is water, most of them land in the ocean!

The chances of being hit are remote. But it has happened! Mrs. E. Hulitt Hodges of Alabama was taken to hospital in 1954 after being hit by a meteorite!

A periodic meteor shower as observed off Cape Florida, November, 1799.

The Cosmic Dustbin

Our planet is pockmarked from the blows of millions of tons of meteorites. And some scientists believe these haphazard visitors from space have had a key effect on shaping the earth we know today:

* The continents of Earth were formed billions of years ago by huge meteorites, according to Klaus Jurgen Schultz, of the McDonnell Center for Space Sciences in St. Louis. He believes a giant meteorite crashed into what is now central Canada, creating a huge bull's-eye ripple effect. The meteorite was 60 to 80 miles in diameter and set off a chain of volcanic eruptions lasting millions of years and which formed the North American continent!

The world's largest meteor crater, Ungava, northern Quebec.

The Tunguska Mystery

In June 1908 there was a massive explosion in the remote Tunguska forest area of Siberia. The blast about four miles above the earth leveled trees over several thousand square miles. Horses were knocked to the ground more than 400 miles away. The night sky glowed for weeks afterwards.

Soviet scientists believe it was caused by a meteor weighing more than five million tons, which caused an explosion equivalent to 300 million tons of TNT!

Others disagree. They say it was a chunk of ice and dust that broke off Encke's comet, which circles the earth every 3½ years.

Other theories:

* A rock of antimatter from another galaxy blew up on contact with the earth's atoms.
* A black hole hit Siberia and passed through the earth.
* It was a nuclear blast, possibly caused when an alien spaceship blew up.

...With a Bang

The earth is in deadly peril from huge asteroids zooming through space, according to some scientists. Depending on its size, an asteroid crashing in the ocean could cause a tidal wave miles high, and even wipe out most of the world's population.

"They could be far more devastating than a nuclear war," says Princeton University Professor Ted Taylor. "The big asteroids hit the earth about every million years or so. If one hit, the effects would be catastrophic."

As recently as 1937, an asteroid named Hermes, about a quarter of a mile across, came within 485,000 miles of Earth—a near miss in cosmic terms!

Rocks from Heaven

Tiny diamonds produced by a collision of two asteroids in space millions of years ago have been found in a meteorite recovered in Antarctica.

Four unusual meteorites found in Antarctica are thought to be pieces of the planet Mars, blasted free in a giant explosion.

Fragments of a meteorite which crashed in Dundee, New Brunswick, Canada, in July 1972, gave off a strange green glow for several days!

Amino acids—the chemical building blocks of life—have been found in two meteorites billions of years old, indicating that the seeds of life are scattered throughout the solar system!

Stroke of Luck

Meteorite weighing 20,000 kg., found by Adolf Nordenskiold (1832-1901) in Greenland.

In May 1698, King Louis XIV of France ordered a road to be built through Great Bayard Rock, near Dinant in Belgium. The day before work was to start, a meteorite split the rock, creating a natural roadway that's still used today!

Raining Cats and Dogs

Rain of toads recorded in 1345.

Nature has a habit of scooping up some extraordinary things—and dumping them many miles away in the midst of a rainstorm! All sorts of things have been lifted up to the heavens by swirling updrafts of air:

* It rained knitting needles in Harrodsburg, Kentucky, in March 1856. The theory is that a needle factory was destroyed in a cyclone—which then died out over Harrodsburg.

* It rained thousands of birds on May 15, 1942, on the town of Pageland, South Carolina.

* A shower of crystallized candy fell from the skies in Napa County, California, in the fall of 1857.

* It rained bullfrogs on Memphis, Tennessee, on October 25, 1946.

* Beads of all colors and sizes cover the fields of Bijori in India in the rainy season. The natives make them into necklaces.

* It rained salt water on Martha's Vineyard, Massachusetts, on August 19, 1896, and on Sunset, Utah, in April 1919.

* Mud drops fell on Tulsa, Oklahoma, on May 6, 1930, and on Abernathy, Texas, in 1935.

* Red rain fell on the Italian Riviera on March 23, 1937—the result of clouds of red dust from Africa swept over the Mediterranean.

* It rains fish every July on Yoro, Honduras.

* A shower of tiny frogs on Trowbridge, Wiltshire, was reported in the summer of 1939.

* Electric rain fell on Cordova, Spain, in 1892. Each drop made a cracking sound as it hit the ground and gave off a tiny spark!

* Periwinkles and small crabs were dumped on Worcester, England, on May 28, 1881, by a violent thunderstorm. Red worms dropped on Halmstead, Sweden, on January 3, 1924; brown worms fell on Clifton, Indiana, on February 18, 1892; and jellyfish descended from the heavens in Frankston, New South Wales, Australia, in October 1935!

Sometimes the problem is simply too much rain:

* The week-long Vermont State Fair held in White River Junction was plagued by rain every single day it was held, throughout its 30 years of existence! The opening date was changed in a bid to get good weather—but in 1928 the fair was abandoned.

Snow Joke

Mother Nature likes her little joke. Just consider some of the strange times and places she chooses for a snowfall:

* A snowstorm hit the Sahara Desert in February 1979. It was so heavy it halted traffic in one town in southern Algeria.
* Snow fell for the first time in 3,000 years on Serir Benaffen and Gargaff in the most arid part of the Sahara Desert on January 6, 1913. Four inches of snow turned the hot sands into a grotesque polar landscape!
* There was an underground snowstorm in a coal mine in Truesdale, Pennsylvania, on February 29, 1934. It was caused by cool, dry air meeting warm air.
* It was a hot day in August 1934 when snow flakes began to fall on a Baltic seaside resort near Godynia, Poland. The reason? Nearby oil wells were pumping enormous amounts of carbon dioxide into the air, which froze the moisture in the atmosphere.
* Worcester, New York, had *two* heavy snowfalls on July 15, 1876.

All Hail

Ever been caught in a hailstorm? We just hope you're never unlucky enough to be hammered by any of these monsters:

* Hailstones weighing from 20 to 108 pounds fell in Yuhsien, China, in August 1933.
* Hailstones the size of oranges battered Seaford, England, on May 30, 1897.
* A storm over Fort Collins, Colorado, on July 30, 1970, covered the area with hailstones half the size of grapefruit. Damage was put at $20 million.
* Huge hailstones smashed down on Moradabad, India, on April 30, 1888—killing 230 people in one day!

* A hailstone six inches by eight inches fell near Vicksburg, Mississippi, on May 11, 1894. Inside was a frozen gopher turtle!

CAKE OF ICE
— ON A WINDOW-SILL
STARTED A FIRE !
—IT FOCUSED THE SUN RAYS
ON A CHAIR AND SET FIRE TO IT.

Theres Ensman of Obersulzbach, Austria, while crossing the Venediger Pass in March, 1935, with her husband and two helpers, was caught by an avalanche which instantly killed the three men. Mrs. Ensman was eventually rescued by her barking St. Bernard, after having been completely enwrapped in the above gigantic snowball for over 24 hours. Believe It or Not!

The Hand of God

THE CHURCH THAT WAS STRUCK BY LIGHTNING TWICE ON THE SAME DAY!
The Petri Church of Berlin, Germany, COMPLETED IN 1726, BURNED TO THE GROUND ON THE VERY FIRST DAY IT OPENED ITS DOORS— IT WAS HIT BY LIGHTNING IN THE MORNING AND THE FIRE WAS EXTINGUISHED— **BUT THAT AFTERNOON IT WAS STRUCK AGAIN AND DESTROYED**

Bolts from the Blue

Does lightning ever strike the same place twice? It most certainly does—and it's not as rare as you might think!

For example:

* The Empire State Building in New York City was hit 68 times in its first 10 years.
* Roy Sullivan, a former Shenandoah ranger, has been hit by lightning no less than seven times! He's been thrown out of his car by a lightning bolt, lost an eyebrow and a toenail, been knocked unconscious and had his hair set on fire.
* The tomb of cowboy star Buck Taylor in Jefferson, Louisiana, has been struck by lightning twice. And he himself was killed by lightning!
* The home of Mr. and Mrs. Perry Owen in Ashland, Alabama, was struck by lightning four times between July 1944 and June 1952.
* Edward Malhern of Chicago was hit by lightning twice in eight hours in 1901—and survived. He was 1,600 feet underground— at the bottom of a gold mine!

Then there have been some unbelievable escapes and some bizarre coincidences:
* Mrs. Minerva Bonham of Tory Hill, Ontario, Canada, had all her hairpins pulled out by a bolt of lightning in August 1951!
* Lightning hit the home of Capt. Bolling Williams of Biloxi, Mississippi, in 1952—and opened a jewel box that had been jammed for years!
* Over a century ago, 12 sycamore trees were planted beside the Grace Episcopal Church in Plymouth, North Carolina, and each was named after one of Christ's apostles. The tree called Judas was hit by lightning— and destroyed.
* Lightning set fire to the homes of C. N. Horne of North Green Street and R. H. Horne of South Green Street in the same storm in Marianna, Florida, in May 1951.
* A 120-foot painting of the Roman emperor Nero in the Gardens of Maius in Rome was destroyed by lightning on the very day Nero committed suicide in a villa four miles away.
* When lightning struck a cowshed in England in 1901, it killed the cow nearest the door, missed the second, killed the third, spared the fourth—and so on, right through the shed of 20 cows.
* The vicar had just begun the service when a fireball struck the church in the village of Widecombe, Devon, on October 21, 1638. People were snatched from their pews and hurled against pillars. A man's head "was cloven, his skull rent into three pieces and

his braines, throwne upon the ground whole, and the hair of his head, through the violence of the blow first given him, did sticke fast unto the pillar or wall of the church; so that hee perished there most lamentably." Three other people were killed and 56 injured in a matter of seconds!

* The city of Brescia, Italy, was destroyed and thousands killed when a bolt of lightning struck the Church of St. Nazaire, setting off 100 tons of gunpowder stored in its vaults!

Lightning has wrecked havoc at a number of sporting events:
* In July 1949 lightning struck a baseball diamond in Baker, Florida, tearing a 20-foot hole in the infield. The shortstop and third baseman were killed instantly, the second baseman fatally hurt and 50 spectators injured.
* Two soccer players died and 17 more were hurt when lightning struck a playing field in Brazil in 1952.
* England's famous Ascot racecourse was hit by lightning on July 14, 1955. Two people were killed and 44 injured.

The **TRAVELING** TREE
DURING AN EXCESSIVE RAINSTORM AN APPLE TREE SHIFTED ABOUT AND CAME SLIDING **200** YARDS DOWNHILL – DEMOLISHING A FENCE AND BARN ON THE WAY – AND **STOPPED UPRIGHT**. – AND *CONTINUES TO BEAR FRUIT*.
KESTERSON FARM, Pomeroy, Ohio

Up in the Air

Hurricanes and cyclones have been fated to change the course of history. The crew of Columbus's fleet were planning a mutiny and might actually have carried it out—but for high winds and a heavy sea that scientists believe were caused by a hurricane. They carried on—and discovered America!

And a cyclone led to the founding of a nation. More than half a million East Pakistanis died after a massive cyclone in 1970.

The Pakistani government was indifferent to their plight and they revolted, setting up the independent state of Bangladesh.

A typhoon in March 1889 actually made peace between Germany and the U.S. After the Germans attacked American property in Samoa in March 1889, three U.S. warships were sent to the scene. But a sudden typhoon sank all the German and American ships before any fighting could take place!

THE WOMAN WHO FLEW ON A FEATHER BED!
ELIZA STATE
DURING THE CYCLONE OF 1860 IN MASSAC COUNTY, ILL., WAS LIFTED AS SHE SLEPT ON HER FEATHER BED IN BOAZ AND CARRIED THROUGH THE AIR INTO POPE COUNTY
— LANDING UNHARMED 8 MILES FROM HER HOME!

It's an Ill Wind...

A great storm saved Greece from invasion in the fifth century B.C. Xerxes the Great of Persia had a great army, but the day the fleet was to set out, a violent storm sank over 400 ships with many soldiers on board. The once invincible Persians had been conquered —by Nature.

Rockabye Baby...

After a cyclone swept through Marshfield, Missouri, on April 18, 1880, a baby girl was found sleeping peacefully in the branches of a tall elm. The child was never identified and was later adopted by a local family.

Religious Bias

A hurricane damaged or destroyed 6,923 churches on the U.S. eastern seaboard in 1938. But through some strange chance, all the synagogues and episcopal churches were spared!

Making Hay

A cyclone in Indiana in 1912 carried off a barn with 50 tons of hay in it and seeded a 40-acre field—just as if it had been done by machine!

Playing Her Song

A tornado that struck El Dorado, Kansas, in 1958 sucked a woman through a window, swept her 60 feet through the air and dropped her unharmed next to a broken gramophone record. The song on the record? "Stormy Weather"!

The **EAST AND WEST TORNADOES**

AN EAST TORNADO KNOCKED OFF THE NORTH END OF M.H. BREDEHAEFT'S BARN —— AND EXACTLY ONE WEEK LATER - A WEST TORNADO KNOCKED OFF THE SOUTH END OF THE SAME BARN! Broken Arrow, Okla.

The Day the Earth Moved

Claude Gay and his party fleeing from the crater of the volcano Antuco, Chile, during a sudden eruption.

The areas most prone to earthquakes include some of the most densely populated areas of the world—parts of Japan and the U.S., and the shores of the Mediterranean Sea. And in a report to the U.S. Congress in 1972, it was estimated that over 500 million people could well suffer damage to property in worldwide earthquake-prone areas—and that many of them could be killed!

Controversial U.S. scientist Jeffrey Goodman believes the chances are good that a major catastrophe is imminent. He says we're entering a 20-year cycle of upheavals and disasters —which have already begun, with the first eruption of Mount St. Helens in March 1980.

Reprieved by a Volcano

All the people of St. Pierre were killed when a volcano erupted in May 1902—except one. Raoul Sarteret, a convicted murderer, watched the explosion from his dungeon cell and was blinded by volcanic gases. He was pardoned—and became a respected missionary!

The Great Sacrifice

The Carib Indians of the West Indies believed they would one day be sacrificed to the fire god they worshiped. The prophecy came tragically true when the volcano called La Soufrière on the island of St. Vincent erupted in May 1902 and killed 2,000 people—including most of the Carib tribe!

Looking Ahead

Here are some predictions by the world's leading earthquake scientists:

* There's a 5 to 11 percent chance that New York City will be hit by a major earthquake in the next 40 years, Dr. Yash Aggarwall and Dr. Lyne Sykes of Columbia University reported in 1978.

* Dr. Kerry Stieh, a Cal Tech geologist, says southern California is hit by major quakes an average of every 167 years and sometimes as often as every 55 years. The last major quake there was in 1857 and Stieh concludes that another can be expected within the next 50 years.

* There could be a major eruption in the volcanic mountains on the northwest coast of North America by the end of this century, according to a U.S. geological survey. And Seattle and Portland could be destroyed.

THE VOLCANIC ERUPTION THAT WIPED OUT AN AMERICAN SUMMER!
THE TEMBORO, A VOLCANO IN INDONESIA, ERUPTING IN 1815, CHANGED CLIMATIC CONDITIONS THROUGHOUT THE WORLD GIVING THE UNITED STATES IN 1816 "THE YEAR WITHOUT A SUMMER" FRUIT TREES DID NOT RIPEN IN THE U.S., BIRDS DID NOT CHIRP, GRAIN DID NOT GROW, AND SNOW FELL IN JULY

Places to Avoid

The U.S. President's Office of Science believes a major earthquake is "inevitable." It says 10 metropolitan areas are in the greatest danger. They are:

* San Francisco. If there was a quake during rush hour, the death toll would be 100,000 or more.
* Los Angeles, which, according to top seismologist Dr. Charles Richter, "will undoubtedly suffer a maximum 8.5 magnitude earthquake soon."
* Salt Lake/Ogden, which are sitting on one of the most active seismic belts in the world.
* Puget Sound/Seattle/Vancouver, where strong quakes are frequent.
* Hawaii, which not only has a number of active volcanoes but is often hit by tsunamis —tidal waves that are generated by earthquakes, often thousands of miles away.
* St. Louis/Memphis, also within one of the world's most active seismic zones.
* Anchorage/Fairbanks: there are more earthquakes in Alaska than in all the other 49 states combined.
* Boston, where earthquakes are rare—but powerful.
* Buffalo, which is in an active zone that follows the St. Lawrence River and the south shore of Lake Ontario.
* Charleston, South Carolina, which has infrequent but strong quakes of unknown origin.

The Psychic Touch

If scientists are gloomy about the chances of a major earthquake soon, psychics are more so. Here's what some of them say, as quoted by Dr. Jeffrey Goodman:

Between 1980 and 1985 there will be:
* Repeated upheavals in California, western Canada and Alaska.
* San Diego, Los Angeles and San Francisco will be destroyed.
* The Great Lakes will grow larger, Palm Springs will be under water and the California coastline will be pushed back to Bakersfield, Fresno and Sacramento.
* A major earthquake will destroy part of New York City.

And between 1985 and 1990:
* Major tidal waves and earthquakes will severely damage India and Japan and will also hit Italy, Greece, Turkey, Iraq, Iran, Syria and Israel.
* New York City will be completely broken up.
* There will be major eruptions of Vesuvius in Italy and Mount Pelee on Martinique.

Not Again!

Severe earthquakes devastated Japanese cities on:
* September 1, 827
* September 1, 859
* September 1, 867
* September 1, 1185
* September 1, 1649
* September 1, 1923

The greatest earthquake in Japan's history was on September 1, 1923. Tokyo and Yokohama were leveled and 143,000 people killed.

Doggone!

Trampess, a dog owned by Kenneth Sandford of San Jose, California, has given warning of 15 earthquakes—sometimes by as much as three days!

Caught in the Act

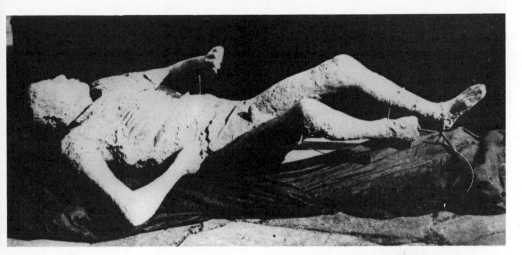

When Vesuvius suddenly erupted on August 24, 79 A.D., dust and ash choked the cities of Pompeii and Herculaneum, but at least 6,000 people refused to leave. About a week later the mountain spewed poisonous gases, and a huge flood of water and lava rushed down the mountainside, mixing with mud and dust. It quickly engulfed the cities, preserving for centuries people and buildings alike.

When excavations started, many perfectly preserved bodies were found, including:

* Gladiators who had just been slain in the arena.

* A man, sword still in hand, with a pile of gold and silver at his feet. Around him lay the bodies of five would-be looters.

* Two Roman soldiers locked in the stocks; and a Roman soldier on guard at the main gate.

* Two servants, complete with keys, money and silver vases, who had just looted a villa.

* A dog, tied to a post and snapping at the chain.

How's That for an Encore?

The great tenor Enrico Caruso thanked his lucky stars that he was away from his native Naples when Vesuvius erupted in 1905 and 1906. But on April 19, 1906, he was caught in the great San Francisco earthquake and ran through the streets, a towel around his throat, clutching a picture of Teddy Roosevelt!

Surviving an Earthquake

Here are some tips on how to boost your prospects of survival if the earth erupts:

* Stay calm, and stay indoors away from windows and heavy furniture. Choose the center of a building and take cover under a heavy table or strong doorway.

* If outside, stay in the open away from buildings and utility wires.

* Do not use matches or candles or other open flames.

* When the shaking stops, check for fires and shut off all utilities.

* Turn on the radio for emergency bulletins.

* Get ready for aftershocks and stay out of heavily damaged buildings.

And now the good news.

The earth quakes about a million times a year, but most tremors are so slight they can't be felt. Only five out of the million actually kill people or damage property.

Web of Fate

The fiery explosion of the Indonesian island of Krakatoa in 1883 was heard as far away as Texas. The first landing party to reach the ruined island found nothing alive—except one red spider spinning its web!

A Forgotten Tragedy

Most people have heard of the Great Fire of Chicago, which broke out on October 8, 1871, and swept through the city, destroying 18,000 buildings and killing 250 people. But few know of a still more devastating fire in Peshtigo, just 250 miles north of the Windy City, because the two fires started on the same day—almost at the same hour!

The Peshtigo fire is the biggest in American history. Of the town's 2,000 people, only 150 escaped death or serious injury. By pure luck, they were protected by an overhanging river bank!

Every building in the town was destroyed in the sudden fire storm that exploded when a forest fire raced down upon the town. The heat was incredible; houses, trees, barns and grass caught fire as if doused in gasoline. People burst into flames as they ran for the river. Of 50 people in a brick house, nothing remained but a pile of white ash and two watches stopped at 10 o'clock. On one spot where a dozen men tried to dig a fire break, nothing was found except their shovel blades.

Ring of Fire

A short circuit set fire to James A. Dewitt's house on Sarasota Bay, Florida, on March 27, 1956—then saved his life by ringing the doorbell to wake him up!

In the Nick of Time

Olle Hult i Fegen, of Halland, Sweden, saw his sawmill struck by lightning and destroyed by fire—less than a minute after signing his first fire insurance policy!

Miracle Baby

The town of Johnstown, Pennsylvania, was wiped out and at least 2,500 killed when a giant dam burst in 1889, sending five billion gallons of water rushing down the valley. But a 5-month-old baby sailed to safety on the floorboards of a ruined home—and was rescued in Pittsburgh, 75 miles away!

The Great Fire of London, 1666.

Disaster in Store

More than 100 Japanese died when fire raged through the Taiyo department store on November 30, 1973. The sprinkler system wasn't working that day—it was being overhauled for fire prevention week!

A Bad Move

With fire sweeping towards his store in Caryville, Tennessee, Charles Duncan carried his stock and furniture outside. Everything he carried out was burned when sparks landed on them—but the building was undamaged!

WATER BURNT A HOLE IN A FIRE HOSE

FRICTION CAUSED BY WATER PUMPED AT THE RATE OF **1,339** GALLONS PER MINUTE *ACTUALLY BURNT A HOLE IN THE HOSE*.

City of Pueblo Fire Dep't. Colorado

Double Trouble

THE HOME OF A. D. JORDEN
Hamburg, Penn.
TWICE BURNED TO THE GROUND

The home of Mr. and Mrs. A. D. Jordan in Harrisburg, Pennsylvania, burned to the ground twice in three years! Both times:
* The house was on the same site.
* The fire was on a Saturday.
* Each was reported by the same person.
* The same fire company rushed to the scene with the same apparatus.
* The cause of the fire was unknown.
* The Jordans had gone shopping.
* The only thing of value saved was the same sewing machine.

Gentle Giant

A tsunami, or mighty sea wave, that hit the Hawaiian island of Oahu in 1946 lifted a house off its foundations, swept it 200 feet inland and dumped it in a cane field. Breakfast was still simmering on the stove!

Surfers' Delight

The highest tidal wave ever recorded occurred in 1737. It was 210 feet high!

Pills Ahoy!

A bottle of nerve pills, swept away when floods destroyed the home of Mrs. Lena McCovey on the Klamath River in December 1964, was found 200 miles away at Coos Bay, Oregon—by Mrs. McCovey's sister!

Oiler on Troubled Waters

Frank Tower, an oiler, swam away from no fewer than three major sea disasters—the sinking of the *Titanic* in 1912, the *Empress of Ireland* in 1914, and the *Lusitania* in 1915.

Porpoise with a Purpose

Hatteras Jack, a white porpoise, guided every ship in and out of Hatteras Inlet off the coast of North Carolina for 20 years—and never lost a single vessel. It would swim around each ship to gauge its size and draw, wait till the tide had reached the right level, then lead the ship safely past the treacherous shoals and reefs. He disappeared in 1810 when buoys and bells made his help unnecessary.

Ship Rock Island in the Bering Sea appeared suddenly amidst the Aleutian islands in 1768—and 130 years later, vanished as mysteriously as it had erupted.

Whale of an Escape

Two Maori women, the only survivors when a canoe sank in Cook Strait, New Zealand, in 1834, were saved by the carcass of a whale. They clung to a harpoon imbedded in the whale and floated more than 80 miles to land.

Kelp Help

The *Ananuac*, a spice-laden clipper sinking off the Cape of Good Hope in a violent storm in 1889, was saved—by seaweed. Capt. Charles Wheldon was about to abandon the sinking ship when drifting seaweed worked itself so firmly into the crevices of the leaking ship that it stopped the leak completely. And 118 days later, the ship sailed safely into Philadelphia.

Saved—by the Bill

Count Felix von Luckner of Germany was a cabin boy on the Russian sailing ship *Niobe* when he fell overboard. He was drowning when an albatross swooped down on him and seized his hand in its beak. The boy hung on to the bird for dear life—and the struggle was seen by the ship's lifeboat which plucked him out of the water.

The UNSINKABLE "HUGH WILLIAMS"

STRANGEST OF ALL COINCIDENCES IN THE HISTORY OF THE HIGH SEAS!

ON DECEMBER 5, 1664, A BOAT SANK IN THE IRISH SEA — OF THE 81 PASSENGERS ONLY ONE ESCAPED - HUGH WILLIAMS!
121 YEARS LATER ANOTHER BOAT SANK IN THE SAME SPOT AND OF 60 PASSENGERS ONLY ONE ESCAPED - HUGH WILLIAMS!
35 YEARS LATER ANOTHER BOAT WAS WRECKED IN THE SAME SPOT - OF 25 PASSENGERS ONLY ONE SURVIVED - HUGH WILLIAMS!

Bottle to Treasure

A bottle, which contained a note describing the fatal injury of Chunosuke Matsuyama and the death of 44 shipmates on a hunt for buried treasure in 1784, was washed ashore 151 years later—at Matsuyama's own village in Japan!

Déjà Vu

Captain Brisco, skipper of the *Grace Harwar*, was washed overboard by a giant wave en route from Delagoa Bay, East Africa, to Gisborne, New Zealand. But he was saved from certain death when a second wave flung him back to his original position on the ship's bridge!

The Human Cork

Casimir Polemus, of Ploermel, France, was involved in three shipwrecks between 1875 and 1882—and each time was the sole survivor!

What a Wreck

The *George F. Whitney*, a schooner launched in 1871, went onto the rocks on its maiden voyage. It was rebuilt and wrecked again on its second voyage. Rebuilt yet again, it sailed into a storm on Lake Michigan—and vanished with all hands.

Tale of Two Othelos

The *Othelo*, the first official lifeboat of Barfleur, France, went out on its first rescue mission on December 4, 1872, and rescued 28 members of an American ship named—the *Othelo*!

Drawn to It!

A drawing of the ship *Seestern* was dropped into the Atlantic in a bottle north of the equator on Christmas Day, 1882, by a seaman named Sachau. And 20 years later, Sachau saw his drawing again—in a native hut on the island of Barbados!

Wrath of the Mountains

It can be triggered quite by chance…a sudden loud noise, thunder or even the vibrations of footsteps. There's a sound like heavy cloth ripping—and suddenly 10,000 tons of snow are roaring down a mountainside at the speed of an express train.

Avalanches are one of the most awesome acts of nature. Some people are caught in them and live to tell the tale. Others are not so lucky:

* 27 people were killed instantly when an avalanche hit the village of Biel in Switzerland in 1869. One family was hurled into the streets while still in their beds; the four children survived but the parents were killed. One boy was sent shooting out of a second-story window—but escaped without a scratch.
* The avalanche that swept down the north face of Mount St. Helens in May 1980 was large enough that it could have buried downtown Portland, Oregon, 40 stories deep!
* A whole forest was carried from one side of a valley at Calancathal, in the Swiss Alps, in 1806. Every tree remained standing!
* An avalanche near Aloerthal, Switzerland, in March 1924 uprooted a chalet and rolled it several miles with a number of children inside. The frantic parents chopped their way in—and found all the children safe!
* British mountaineer Dougal Haston, one of the two men who were first to scale Mount Everest by the southwest face in 1975, was swept to his death by an avalanche while skiing in the Alps in January 1977.
* The worst avalanche in U.S. history happened in the Cascade Mountains. Washington, in 1910. A wall of snow threw several trains—and the train station—into a canyon 150 feet deep. A hundred people were killed.
* Over 10,000 soldiers were killed by sudden avalanches in the Tyrol section of the Alps in one 24-hour period in World War I! Troops had shelled the snowfields above the enemy to start massive slides. Between 1915 and 1918, an estimated 60,000 soldiers died in the freezing Alps!

THE CHURCH THAT WAS DESTROYED BY ITS OWN BELLS
Saint Nicolas , Switzerland
THE BELLS in the belfry of the **CHURCH OF ST. NICOLAS**
SET OFF A SNOW AVALANCHE THAT SWEPT AWAY
THE CHURCH – YET THE BELL TOWER ITSELF
WAS UNHARMED
*THE BELL RINGER WAS NOT EVEN AWARE THE
CHURCH HAD DISAPPEARED UNTIL HE DESCENDED
FROM THE ADJACENT TOWER* (1749)

Doing What Comes Naturally

Mother Nature continues to surprise us, in the most unexpected ways. Consider these chance acts:

* A large maple tree cut down by E. J. Webb, Jr., of Statesville, North Carolina, had a wild cherry tree 20 feet high growing inside its hollow trunk!
* A lemon weighing 22 ounces was grown on a dwarf lemon tree in 1954 by Mrs. Ernest Borchardt of Leavenworth, Kansas.
* The thorn of Dagny in France, said to have grown from the staff of St. Geroche when he thrust it into the ground in 663 A.D., has bloomed every Christmas for over 1,300 years.
* The Manchurian lotus in the Kenilworth Aquatic Gardens, Washington, D.C., was grown from a seed that had lain dormant in the bed of a lake in Manchuria for thousands of years.
* A cherry tree planted in Grand Rapids, Michigan, as a single seedling produced both cherries and peaches.

Inventions: A Lucky Break

Where would the world be today if the telephone had never been invented? Or X-rays? Or even the lowly hotdog? Hundreds of major scientific advances have been brought about by serendipity—looking for one thing and finding another quite by chance.

American scientist and sci-fi writer Isaac Asimov puts it this way: "Since important discoveries are invariably unexpected (if they were expected, they would be too routine to be important), they are almost always serendipitous. They have to be."

Here are just some of the miraculous discoveries that have brought joy—and sometimes misery—to the whole of mankind.

The Telephone: That Long-Distance Feeling

Love it or hate it, the telephone is here to stay. But it all came about by accident!

Alexander Graham Bell and his assistant were working in a garret in Court Street, Boston, on a machine to help the deaf. On the night of June 2, 1875, Watson was working on the reeds of a harmonic telegraph, and Bell was on the other end of the line in another room. A reed had been screwed down too tightly and Watson plucked at it to free it. An excited Bell heard a twang on the other end of the line.

Then after much tedious experimenting, Bell and Watson were trying out a new liquid transmitter in Exeter Place, Boston, on March 10, 1876. They were in different rooms when Watson suddenly heard the words "Mr. Watson, come here. I want you!" Bell had spilled battery acid on his trousers and a few seconds later Watson burst into the room, saying, "Mr. Bell, I distinctly heard every word you said."

Bell's words were the first ever uttered over wires through which an electric current flowed. And the rest, as they say, is history.

A Bad Line

The first telephone circuits were so bad that the word *phoney* meant "as vague and unreal as a telephone conversation."

Man of Destiny

Kitchen Battlefield

German chemist C. F. Schobein disobeyed his wife in 1845—with explosive results!

She'd told him to keep his smelly chemicals out of her kitchen but he sneaked in while she was out shopping. He spilled nitric acid on the floor, mopped it up with her apron and hung it up to dry over the stove. The result? An explosion! The cellulose in the apron combined with the acid to make nitrocellulose—an explosive that's now replaced gunpowder on the field of battle.

A Bright Idea

German scientist William Roentgen was experimenting with cathode rays in 1895—when he discovered X-rays! There happened to be some paper coated with barium platinocyanide nearby, which glowed when he turned his machine on. He put his hand between the apparatus and the screen—and saw the outline of his bones!

A New High

Swiss chemist Albert Hofmann was investigating a new chemical in 1943 when he absorbed some through his skin and began to hallucinate. He later investigated the strange substance—and discovered LSD.

A Pressing Issue

Someone knocked over a lamp in the home of a French dyemaker in 1825. The fuel removed the stains on the tablecloth...and dry cleaning was invented!

The First Papermakers

We owe the invention of modern paper to a sharp-eyed French naturalist.

Entomologist Rene Antoine Ferchault de Reaumur was walking through the woods in the early days of the 18th century when he passed a wasps' nest. He examined it and found it was made of a crude form of paper. He concluded that the wasps had digested tiny bits of wood to produce the paper—and the modern process of making paper from wood pulp was born!

Man of Destiny

It would be hard to imagine a more unlikely man to invent the telegraph. Samuel Morse was an American portrait painter who had practically no knowledge of electricity, yet he was destined to revolutionize electrical science with his invention. And yes, it was pure chance that he did so.

His wife died in 1929 and the grief-stricken Morse decided to visit the art galleries of Europe. On board the packet *Scully* on his way home, he heard a lecture from the prominent American scientist Dr. Jackson, who remarked, "The electric current is instantaneous."

Morse remembered that one of his college teachers had said electric current was fluid. The current could be interrupted, he reasoned, and clicks could be made by an electromagnet opening and closing the circuit—a short click for a dot, a long click for a dash.

Finally the first telegraph line was set up between Baltimore and Washington, and Morse sent the first telegraph message in May 1844. It was: "What hath God wrought?"

A Tired Argument

Charles Goodyear had an argument with his brother and discovered vulcanized rubber! Gesturing wildly, he dropped a chunk of rubber gum and a piece of sulfur onto the hot stove behind him. The result was an elastic, stable and tough material, with countless possible uses.

But Goodyear didn't profit from his discovery. The process was easily copied and he died in debt before vulcanized rubber became widely used.

Knocked on the Head

It's one of the most famous stories about chance and scientific breakthroughs. Gravity was discovered, so legend has it, when Sir Isaac Newton was hit on the head by an apple while sitting under a tree. Alas, the story may not be true. Newton had been working on the idea for years. But it was a stunning practical demonstration of his theory!

Wiping out Calories

Student chemist Constantine Fahlberg discovered saccharin—all because he wiped his mouth with his hand. The new compound on his fingers tasted sweet. It was saccharin!

You've Heard This One Before

Legend has it that Greek mathematician Archimedes discovered the principle of specific gravity while taking a bath. He noticed that the water went up when he lowered himself into the bath. "Eureka," he cried and raced home naked to try measuring the volume of objects by seeing how much water they displaced. The technique is still used today!

A Load of Hot Air

Frenchman Joseph Montgolfier was watching smoke go up his chimney in 1782—and invented the hot-air balloon!

He figured out that hot air expands and rises because it is lighter than cool air. He built himself a paper-lined linen balloon in 1783, put charcoal fires underneath—and the idea took off!

The Mysterious East: A Good Connection

Hans Christian Oersted, a Danish professor of electricity and physics, was in the midst of a lecture in January 1819 when he discovered electromagnetism!

He was demonstrating a wet-cell battery when he noticed that a nearby compass was pointing east instead of to the magnetic north. And presto, the connection between electricity and magnetism was made!

The Idea That Clicked

"George," a friend said to young bank clerk George Eastman as he was about to take a vacation, "why don't you take a camera along and make a record in pictures of your trip?"

The young bank clerk was fascinated by the chance remark. But the 1870s were still the early days of photography and the problem was daunting. It would have meant carting along a huge camera, heavy tripod and chemicals and glass plates. So Eastman thought up a method of replacing glass plates with rolls of paper coated in chemicals and "cranking" the frames into place in a box camera. And that was the start of all those holiday snapshots!

Inventions: What a Crazy Idea

Got a better idea? Chances are that the world may not quite be ready for it yet. Consider some of these discoveries:

* The photocopier was invented by Hungarian Andor Rott in 1939. Chester Carlson, an American, patented his version in 1948—but the idea didn't gain widespread acceptance until Xerox and IBM started manufacturing the dry copier cheaply in the 1960s.
* The incubator was invented by an ingenious physician in 1577 to save his premature baby, who lived to be 80. Yet incubators only came into general use in hospitals this century.
* Microfilm was used during the siege of Paris in 1870 when minute copies of the London *Times* were flown in by carrier pigeon. Yet the invention languished until the 1940s when it was rediscovered.
* The first zip fastener was invented by Whitcomb L. Judson of Chicago in 1891. The modern-day zipper of identical units mounted on parallel tapes was patented in April 1913 by Gideon Sundback, a young Swedish engineer from Hoboken, New Jersey. But it was not until 1930 that the first zippers appeared on Paris dresses and 1935 that zippers on men's trousers were introduced!
* S. C. Gilfillan studied the history of 19 inventions that came into general use between 1888 and 1913. He found it took an average 176 years before the idea was actually tried out, another 24 before it became practical and 14 more before it made any money!

Of course the pace is hotting up these days. But some modern inventions were on the shelf for some time before becoming reality:
* Photography was conceived in 1782 but only came into use in 1838.
* Television was thought of in 1884 but we first began to tune in 63 years later.
* Nuclear energy seemed like a good idea in 1919 but only became reality in 1965.
* Instant coffee was introduced in 1934 but only began to be produced in quantity in 1956.
* But filter cigarettes were invented in 1953 —and were on the market by 1955!

Food for Thought

Just think of it! A world without hotdogs, chips or ice-cream cones! Yet all were fluke inventions:

* The hotdog came about in St. Louis in 1904 because Anton Feuchtwanger, an immigrant from Bavaria, couldn't afford plates for his frankfurters. So he stuck them in a bun instead!

* A row between George Crum and fussy guests led to—potato chips (crisps)! Crum, an American Indian who was a chef at Saratoga Springs, New York, in 1853, was furious when his customers sent back his chips as "too thick." So he sliced them as thin as possible and the clients loved them! In fact, until 1900 the new delicacy was called "Saratoga chips" in his honor.

* The ice-cream cone was invented in St. Louis at the World's Fair in 1904. And it was all because a vendor ran out of dishes! Ernest Hamwi, a Damascus-born wafer salesman, came to the rescue. He twisted hot wafers into a cone shape, let them cool, and the scoop of ice cream was ladled on top. An instant success!

* The cocktail was born in a small tavern in Westchester County, New York, during the American Revolution. Barmaid Betsy Flanagan stirred a mixed drink with feathers from a rooster's tail. A Frenchman shouted, "Vive le coq's tail." Et voila!

Poor Ideas

Will your invention make you rich and famous? Probably not:

* Only 1 percent of all patents ever become successful products.

* Only 10 out of 150 ideas are ever developed. Of these only two are marketed and only one succeeds.

* More than 1,800 patents are granted every week in the U.S. Only 3 percent are marketable.

What the Future Holds in Store Computers: Terminal Man

Love 'em or hate 'em, computers are here to stay. In the not-too-distant future, we'll probably all have our very own home computer terminals, plugging us in to a storehouse of information.

* By 1990, your doctor may be a computer, questioning you about your health and coming up with a complete diagnosis. In fact, patients who have been examined by computers have given them more candid answers than they would have given their doctors!

* Computers may mean a cashless society. Computer banks will simply issue you credit from any one of hundreds of terminals, probably across the world. In fact, this new form of bank already exists in many countries.

* Talking computers will run households and businesses as well as book your holiday and provide you with all sorts of video amusements.

* Whole libraries will be available at the touch of a button, as will the morning newspaper, stock market prices and a host of other information. You will be able to shop from home and order things from your favorite supermarket. Your home video terminal may even let you work at home!

Spirited Try

One thing computers have trouble with is language; they can't really understand the meaning of words. One computer, told to translate "The spirit is willing but the flesh is weak" into Russian, came up with "The vodka is good, but the meat is rotten"!

The Future: A World of Wonders

George and Edward Scheutz's calculating machine, developed between 1837 and 1843.

The microchip and other marvels of modern technology are about to change the way we live. Here are just some of the things that could revolutionize things in the not-too-distant future:

* Books and newspapers as we know them will disappear, to be replaced by video disks. You'll be able to call up your favorite novel or magazine from a computer bank in a central warehouse and project it onto your own video screen at home.

* Your car will get around 100 miles per gallon—just as well since gas will be $8 a gallon or more! Or it could be running on a mixture of gasoline and alcohol called "gasohol" or on propane.

* Vacuum tubes may be able to whisk you from New York to Los Angeles in just 21 minutes; other systems would speed around the world at about 14,000 miles an hour.

* Moving sidewalks will replace stationary concrete ones—they're already in use in several major airports and Osaka's Expo '70 used them to move over a million people a day.

* Airports may be built on island modules with runways that are extendable to any length—all you have to do is add on more modules. And they would be hollow, with space for offices, apartments and hotels.

* The cities of the future could be enclosed in domes—ultraefficient and climate-controlled.

* Domes could be used to turn the South Pole into farmland. Futurist Buckminster Fuller believes the thick ice-cap could be melted, exposing some of the most fertile land on earth.

* The age of sail may not be dead. Oil tankers of the future could use sails. The Japanese have already tested a sail-assisted tanker, the *Shin Aitoku Maru*, and reckon the computer-trimmed sails will cut fuel costs in half.

March of the Robots

Could a robot do your job? Chances are that it could. It may even do it *better* than you can, particularly if you are in the manufacturing industry!

Experts at Pittsburgh's Carnegie-Mellon University reckon that eight million of America's 10 million workers in manufacturing will be replaced by robots within two generations!

Robots are:
* Faster and better welders than humans.
* Stronger and tireless packers.
* Impervious to heat and fumes in foundries.
* Work for an average cost of 80¢ an hour, $10 less than their human counterparts.
* Don't strike, make wage demands, want overtime pay, need sick benefits.

They're All around Us

There are about 4,000 relatively sophisticated robots at work in the U.S., mainly in the auto industry. Japan has an estimated 70,000 and there are several hundred in Europe. And they're becoming more and more intelligent. "A generation of robots is rapidly evolving, a breed that can see, read, talk, learn and even feel emotions," says cyberneticist D. Rorvik.

Flash Food

The chances are that the cook of the future will be using a microwave oven. Here are some predictions by Anchor Hocking Products of Minnesota:

* 30 percent of Japanese have a microwave now; by 1990, 50 percent will have one.
* 20 percent of Americans own a microwave; by 1985, 40 to 50 percent will.
* 10 percent of Canadians use a microwave now; 25 percent will have one by 1986.

Necessity Is the Mother of...

The idea came to French-born inventor Michel Deal as he struggled up a long hill outside Bath, England: Why is there no automatic transmission on a bike?

He perfected the idea in Montreal and it won first prize in the annual Lepine Competition of Invention in Paris in 1980. It will be available soon.

Back to the Drawing Board

Just a little slip of the pen can produce some extraordinary bloopers. Such as:
* Crews working on a Pennsylvania bridge started from opposite banks and worked towards the middle. Work came to a halt when they discovered they would miss each other—by 15 feet!
* The Chinese spent years copying the American Boeing 707 jet but forgot just one thing: the center of gravity. Now their expensive copy, the Y-10, can't fly.
* On November 4, 1981, the U.S. space shuttle failed yet again to take off. Somebody had forgotten to check the oil!

Love That Computer

London factory worker William Waldron became so upset when his favorite computer was taken away for repairs that he insisted on staying at home—without pay—until it came back. After 10 months the computer still hadn't come back and William was fired. But in October 1981 a court awarded him £3,500 for unfair dismissal and almost £1,500 in severance pay!

Turning on Its Creator

Japanese factory maintenance worker Kenji Urada was crushed to death by a robot in July 1981, the first recorded death blamed on a robot!

Clyde's Nerves

A computer called Clyde the Claw was working in an auto plant when it suffered something approaching a nervous breakdown. It pulled in its arm and refused to work! So fellow (human) workers had a party and sent him flowers and get-well cards. Before long Clyde was better!

Shell-shocked but Undamaged

A trailer truck carrying 400 crates of eggs hit the Rhyne Bridge at Charlotte, North Carolina, in 1956. Damage to the truck and bridge was put at $6,000—but not a single egg was broken.

They Kept Running into Each Other

Cars driven by Miss Joan Jones of Mountain Lakes, New Jersey, and Earl W. McComan of Washington, D.C., collided on Lake Street in Penn Yan, New York, in 1959. One year later two cars collided at the same spot—driven by the same two drivers!

Twin Mishap

Dwight Peterson of Lafayette, Colorado, lost control of his car and shot over an embankment in March 1964. Moments later and right behind him, his identical twin lost control of his car—and drove over the embankment at the same spot!

I'm Thirsty, Daddy!

John and Lynette Roth of Oxnard, California, have wired their 22-acre avocado grove up to a computer that checks soil conditions and waters the plants when they are thirsty. In an emergency, an avocado tree can wake them up at three in the morning at their home 15 miles away—and ask for a drink!

Chance Train of Events

"One-Eyed" Jerry Simpson was working on a bridge in the Cascade Range, Washington state, in the fall of 1886 when a runaway North Pacific engine came hurtling towards him. Preferring instant death to crippling injury, Simpson threw himself across one of the rails. But at that very moment, one side of the engine lifted from the rails and passed over him, crashing into the gully below!

Like Father...

Carl A. Waldman of Reading, Pennsylvania, was hit by a truck at the corner of 11th and Spring streets in Reading on May 14, 1970— at exactly the same time as his son Michael was injured by a car at the corner of 10th and Spring. They arrived at the same hospital at the same time—but neither was seriously hurt!

He Bearly Escaped

Mates Huberli, a woodchopper in Lebern, Switzerland, was attacked by a bear at the edge of the great precipice of Wandfluh. Locked in the bear's embrace, he plunged 4,000 feet—but lived when he landed on top of the bear! The bear was killed.

Satcom 3, Where Are You?

Whatever happened to *Satcom 3*? The satellite, the size of a household refrigerator, weighed 907 kilos and had seven-meter wings. After a perfect NASA launch, it simply disappeared!

Where is it now? There are various theories:

* It just went shooting off into space.

* Its transmitter is broken, but it's still up there—somewhere.

* It fell back to earth and blew up.

Strange Cultures: Alive, Alive-o

A lie detector expert says he's detected electrical impulses passing between two containers of yogurt at opposite ends of a room and concluded they were talking to each other.

A neurologist in Ontario, Canada, attached electrodes to lime jelly and detected signs of life!

Space Food: Compliments to the Chef

Would man be able to eat in space? During the early days of space missions scientists weren't sure—they thought zero gravity might make it impossible!

Space food has come a long way since then:

* In 1962, John Glenn had to squeeze his food out of something resembling a toothpaste tube. It had no smell and no taste, but started a craze among kids, who squeezed treats from tubes.

* By 1968, astronauts Borman, Lovell and Anders were able to feast on turkey, gravy, cranberry sauce and all the trimmings for their Christmas dinner.

* By 1973, astronauts on the Skylab mission were eating warm food, selected from a menu containing such items as prime ribs, lobster Newburg and filet mignon. Space food has become almost indistinguishable from ordinary food—except, perhaps, that it's better!

Man Digs Potatoes!

The humble potato has a checkered history. But its popularity today came about largely by chance:

* First brought to Europe in the 16th century by the Spanish, who had found them being cultivated in South America, potatoes were thought to be an aphrodisiac because they grew sprouts shaped like a human phallus.
* They were feared as a poison because they belong to the same family of plants as the deadly nightshade.
* In Russia the potato was often considered harmful to other crops and superstitious peasants often burned it at the stake to punish it for bad harvests.

* Antoine Auguste Parmentier popularized potatoes in France, after growing to love them in a German prison during the Seven Years' War! He threw lavish dinners for his influential friends at which every course consisted of potatoes served in different ways! He even gave a necklace made of potatoes to Marie Antoinette! The potato dish potage Parmentier is named after him.
* The potato has earned an undeserved reputation as being fattening; full of vitamin C and minerals, it contains only 85 calories. It's the melted butter or the sour cream that can put on the pounds!

PART FIVE:

Tangled Web

Crime and the Scales of Justice

We live in a crime-ridden world. Few of us go through life without stepping outside the law in one way or another—even if it's only breaking the speed limit. Most of us have met criminals, knowingly or not.

Here you'll meet them all, from infamous murderers to incompetent thieves. And it's the whim of fate that decides whether they will pull off the crime of the century or end up on the gallows or in the electric chair.

Read about:

* The men and women they could not hang!
* The great computer crimes.
* The laughable, luckless fools of fate.
* The men who were guilty—until fortune found them innocent.
* The great con men who sold the Eiffel Tower, Big Ben and Buckingham Palace to the gullible.
* The localities in which the law is definitely an ass.

The famous highwayman Dick Turpin (1706–39) riding from London to York on his gallant mare, Black Bess.

Murder Most Foul

Detectives investigate a murder.

Your chances of being murdered are slim—but growing!

A murder is being committed every 24 minutes somewhere in the United States. Numbers rose from 7,990 in 1964 to 18,714 in 1978. Odds against being shot to death? 20,000 to 1. Of 2.5 million crimes committed in Britain in 1979, only 571 were homicides. But that was 50 percent more than it was 10 years ago.

Born to Die

A child born today has a greater chance of being murdered than a U.S. soldier did of being killed in World War II, says futurist Laurin A. Wollin, Jr. And one out of every 61 babies born in New York City in 1980 can expect to die at the hands—or the gun—of a killer, according to a study by the Massachusetts Institute of Technology (MIT).

To the Ends of the Earth

Charles B. Henry, a member of the ill-fated Greely Arctic expedition of 1881, was executed by his fellow explorers for stealing food. The group was stranded in the bitter cold for three years and 15 men died of hunger and exposure.

Later it was discovered that Henry's real name was Charles Henry Buck. He'd escaped from prison while awaiting trial for murder.

The MURDERER WHO DISCOVERED THAT JUSTICE IS **NOT** BLIND!

WILLIAM WILLIAMS of Lick Fort Creek, Alabama

WHO MURDERED HIS FRIEND IN A SIMULATED HUNTING ACCIDENT TWICE ESCAPED FROM PRISON *BUT WAS RECAPTURED WHEN HE WAS STRICKEN BLIND BOTH TIMES – AS SOON AS HE HAD LEFT HIS CELL!*

EACH TIME HE RECOVERED HIS SIGHT AFTER BEING RETURNED TO PRISON

Telltale Fragments

There's a 20-to-1 chance that at least one fiber of clothing or body hair will pass from victim to assailant or vice versa during homicides or other acts of violence. If five such fibers are found on a person, the odds are a staggering 3,000,000 to 1 that the two have *not* been in contact with each other.

You can be saved by a hair. And thanks to modern science, it's becoming less and less of a long shot. For example:

* Neutron analysis can determine exactly what elements are in a person's hair. No two people's hair is alike. Molybdenum, gold, zinc, iodine, chlorine, chromium, selenium, manganese, mercury, copper, cobalt and other elements have all been found in human hair.
* Experts can tell what part of the body a hair came from, whether it's been cut recently, whether it's been bleached or dyed. They can detect arsenic poisoning, since traces of arsenic will be found in the hair, and whether

it was pulled out or came out naturally.
* Age, sex and overall color of hair are more difficult because of the variations involved.

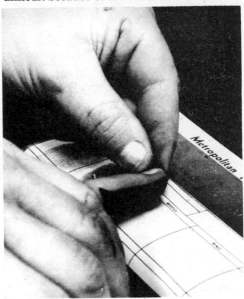

It Pays to Advertise

Jesse Boorn of Manchester, Vermont, was cleared of a murder charge by a classified ad. Convicted of murdering his brother-in-law in May 1812, he was cleared when his brother placed an ad in the Rutland *Herald* and the "murdered" man was found alive and well in Dover, New Jersey.

The Illustrated Man

The criminal least likely to have a tattoo is a murderer, studies in U.S. penitentiaries indicate. Tattoos are most frequent on burglars, then armed robbers, then rapists, kidnappers and would-be murderers.

Golden Dreams

Women in Sipche, Nepal, poisoned all the men in the village in 1973 because they believed the 100th victim would turn to gold and make them all rich.

Charlie Appleseed

Charles Boyington, hanged in 1835 for murdering his friend, predicted an oak tree would spring from his grave to prove his innocence. Contemporary accounts say Boyington was unquestionably guilty—but an oak tree grew out of his grave.

Slippery Customer

Bella Krager, who was just 5 feet 1 inch, was jailed for murdering the curator of the P. T. Barnum Museum—but escaped by buttering his body and slipping between the bars of his cell!

Bang! Bang! You're Dead

Sixty percent of the murders in the U.S. in 1978 were committed with guns, while knives accounted for another 23 percent. And in Britain, where carrying guns is banned, the number of criminals now using firearms is five times what it was 10 years ago.

Hurting the Ones We Love

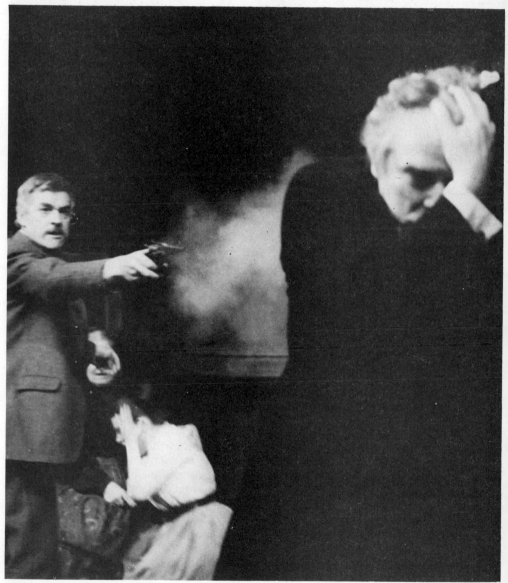

*Murder scene from Anton Chekhov's **Uncle Vanya**.*

Most homicides involve a family or friends. Every 10th murder involves a husband and wife. But in the U.S. one third of all murders are random, done by strangers, often for no reason!

All in the Family: The Mob

The infamous St. Valentine's Day massacre, 1929; "Bugs" Moran's bootlegging gang gunned down in cold blood by Al Capone's henchmen.

The second biggest business in the U.S.? No, it's not the chemical industry, auto manufacturing or steel. It's organized crime.

Figures presented to a Senate subcommittee estimated that it grossed more than $150 billion in 1979, second only to the oil industry with revenues of $365 billion. And much of it is tax-free!

Some facts:

* The Mob made an estimated $63 billion from narcotics in 1979, $22 billion from gambling, $8 billion from assorted crimes such as arson and prostitution, and possibly $20 billion from loansharking.

* Organized crime has probably turned out more millionaires than the Harvard Business School.

* It has moved strongly into legitimate business—Wall Street, real estate and the hotel business, video games and gambling casinos.

* It has about 5,000 criminals working for it across the country.

* The bill for keeping Mob hit-men away from the 2,000 government "squealers" is $20 million a year.

* It got a foothold thanks to two blunders by the U.S. government. In 1890, the government paid $30,000 compensation to the widows of 11 immigrants who had been lynched, money used by the brotherhood for its first organized extortion operation. Prohibition in the 1920s gave the Mob the chance to bootleg liquor and by the end of Prohibition it was well established.

Fortress America

Ninety percent of people in America always lock their doors, 52 percent own guns to protect themselves, 40 percent fear assault, robbery, rape or murder every day, and 50 percent support forced sterilization of habitual criminals and the insane, according to a survey by ATO, an Ohio manufacturer of security systems. The fear of crime seems to be rising faster than crime itself, the survey concludes.

Caught—by Chance

Some of the world's most notorious and bloody criminals have planned their crimes in meticulous detail—only to be caught by human error or an odd quirk of fate.

* George Joseph Smith drowned three of his wives in the bathtub, establishing as his alibi that they were subject to fits—and then collected on their life insurance. But the father of an earlier victim subsequently read a newspaper report of another wife's death, recognized the pattern and told the police. Smith was hanged in Maidstone, England, on August 13, 1915.

* A suspicious ship's captain and wireless telegraphy brought the arrest of American-born murderer Hawley Harvey Crippen. After killing his young nymphomaniac wife, Crippen fled for Canada with his young

A London "Peeler" about to make an arrest.

mistress who was disguised as a boy. But the captain noticed her feminine appearance and the affection between the two. In the first such use of wireless telegraphy, he wired Scotland Yard and the couple were arrested when the ship docked. Crippen was hanged in 1910.

* A $35 parking ticket slapped on a car parked too close to a fire hydrant led New York police to David Berkowitz—the infamous Son of Sam who stalked the streets in 1976 and 1977, killing six people and wounding seven others.

* Peter Sutcliffe, the Yorkshire Ripper who killed and mutilated 13 women in the north of England, was arrested in January 1981—after a young constable found that a car he was checking out had false number plates and arrested the driver.

* Britain's Great Train Robbery in Buckinghamshire in August 1963 was the perfect crime—almost. The robbers made one incredible blunder. After getting away with nearly £2 million, they fled to a remote farmhouse. But an associate failed to clean up after they left and police found damning evidence—fingerprints, clothing and vehicles. Within a year, most of the gang were in jail.

The Yorkshire Ripper

Generally Speaking

Every three seconds a crime is committed in the United States. Every 27 seconds there's a violent crime—murder, rape or assault. Every 10 seconds a house is burgled, every seven minutes a woman somewhere is raped. Crime is a boom industry.

In Britain, muggings have increased 300 percent in the last decade while vandalism is up 200 percent.

Chances of Conviction

If you commit a crime in the U.S. your chances of escape are good:

* Only three out of five criminals are ever caught, even when the crime is reported.
* Only one in 100 computer crimes is ever detected, much less successfully prosecuted.
* One in three people tried is acquitted.

Locked Away from It All

Contemplating a quiet life? Here's where to look:
* One in 3,125 Swedes is in jail;
* One in 3,571 Danes is a prisoner;
* One in 5,555 Netherlanders is in prison;
* One in 465 Americans is in prison.

The Great Northern Stickup

What's the top North American city for bank robberies?

Odds are you'd pick New York or Los Angeles. But no, it's Montreal, Canada with 832 robberies in 1980, compared with New York's 687 and Los Angeles's 387.

Kid Stuff

In any given crime, the likelihood is that it was committed by a child under 15.

More kids under 15 commit crimes than adults over 25, according to the latest U.S. statistics. The number of juvenile crimes has risen an astonishing 1600 percent in the last 25 years.
* Marvin Wolfgang, of the University of Pennsylvania, studied 10,000 boys born in 1948. By the time they were 16 years old, 35 percent had been arrested at least once— then almost all stopped committing crimes.

Identification: Sorry, Wrong Man (or Woman)

You're innocent. But sometimes the long arm of coincidence reaches out and fingers you for a crime that you never committed. You're guilty—until proven innocent!
* Pharmacist David Feinberg was standing at the scene of an arrest for stolen goods when the criminal pointed him out as his supplier. Ironically Feinberg had been convicted for just this offense years before. It took him seven years of his 40-year sentence to prove his innocence.
* Philip Caruso was convicted of armed robbery because of a cold sore on his lip, which a witness identified as a mole. He was released from his 10- to 12-year sentence only when the real robber confessed.
* Mrs. Emma Jo Johnson was convicted of murder after she fought with her 72-year-old landlady, who collapsed and died. In 1954, three years after Mrs. Johnson's conviction, it was proven that the bloodclot that killed the landlady was there before the fight and could have killed her at any time.

A simulated identification line-up with police detectives.

* James W. Preston was arrested and convicted in 1924 of robbery and assault—even though his fingerprints didn't match those of the real robber. The Los Angeles newspapers had carried a false story accusing Preston of being the robber and the victim read it and identified him as her assailant. The only man who thought he was innocent was the fingerprint expert, who 18 months later caught the real culprit.

* William Green was convicted of murder because a homosexual had a grudge against him. On the evidence of two eyewitnesses, Green was jailed for life in 1947 for murdering a night watchman. Ten years later one of the witnesses confessed the other one had paid him £50 to make up the story. The other witness was a homosexual beaten up by Green, a navy veteran, for making an indecent proposal to him.

* Lonnie Jenkins of Detroit was accused of murdering a suicide—on the testimony of a 15-year-old girl who was infatuated with him. Jenkins's wife had shot herself and left a suicide note but police continued investigations because the body was found in an unusual position. The girl told police Jenkins had promised to kill his wife and marry her, even claiming she'd copied the wife's handwriting for Jenkins.

Jenkins was found guilty and sentenced to life imprisonment but his lawyer kept working on the case. He worked out how a person could shoot herself and fall the way Mrs. Jenkins had, went to the police station to demonstrate his findings and accidentally shot himself dead!

Finally, the girl admitted she'd invented the story and Jenkins was freed in 1940—after spending nine years in jail.

Clear as Day

Most witnesses are lucky if they can get a clear view of the criminal. Many are in shock. Sometimes their glasses are broken and often they are shortsighted.

* Arthur Thompson, wrongly convicted of robbery with violence, was identified by someone three stories off the ground, at three in the morning, by the light of a small electric bulb, and through three large trees that completely obscured the view!

> **NICOLAS SHEEHY** (1728-1766)
> of Fethard, Ireland
> WAS UNJUSTLY CONVICTED AND EXECUTED FOR MURDER—*AND ALL 12 MEMBERS OF THE TRIAL JURY SUBSEQUENTLY* **DIED VIOLENT DEATHS**

I Saw Him Do It

Identification is a risky business. An analysis of 20,000 eyewitnesses, quoted by Jerome and Barbara Frank in their survey of American cases of wrongful imprisonment, revealed:
* Height is overestimated by five inches;
* Age is overestimated by eight years;
* The wrong hair color is given in 82 percent of the cases.

An Eye for an Eye

Here's the 18th-century French philosopher Voltaire on eyewitness reports: "He who has heard the same thing told by 12,000 eyewitnesses has only 12,000 probabilities, which are equal to one strong probability, which is far from certain."

Hangman's Bluff

Some people are born to hang.

And some have had incredible escapes when fate relented at the last possible minute:

∗ John Smith of Malton, England, was hanged at Tyburn the day before Christmas in 1705—and lived to tell the tale. Revived after 15 minutes, he became a great raconteur and was nicknamed Half-hanged Smith.

∗ Joseph Samuels of Sydney, Australia, was thrice lucky in 1803. The rope broke three times and officials let him go.

∗ Another three-time winner was John Lee of England. Sentenced to death for murder, he survived three attempts to hang him at Exeter Jail on February 23, 1885. Each time the trapdoor failed to open, and his sentence was changed to life imprisonment. He was released after 22 years and emigrated to America.

∗ Sentenced to hang for murdering a neighbor, Will Purvis of Columbia, Mississippi, cheated death in 1894 when the noose slipped. Later cleared of the charge, he outlived all the jurors who had convicted him.

∗ In 1728 Margaret Dixon was hanged in Scotland for murder, and her body was left dangling on the gallows for a full hour. Yet she revived as her husband was carrying her body to the cemetery—and lived for another 25 years.

∗ William Duell was hanged at Tyburn in 1740, but a surgeon noticed signs of life when his body was being washed. Treated and returned to prison, he was tried again and sent to Australia.

∗ Patrick Redmond of Cork, Ireland, had been hanging on the gallows for 28 minutes in 1767 when a mob carried off his body and he was revived. He was pardoned and a collection was made for his welfare.

∗ Andrew Tracy of Smethport, Pennsylvania, was being hanged for killing his cousin when the rope broke. He plummeted to the stone floor seven feet below the scaffold and died of his injuries.

∗ Thomas Reynolds was hanged for robbery in 1736, cut down and put in his coffin. He suddenly sat up, thrust back the coffin lid and was carried off by an angry mob. He threw up three pints of blood, drank a glass of wine—and fell dead.

Two Two Bad

Arthur Cameron, convict No. *22222* in the Ohio Penitentiary, was arrested by *two* men for stealing *two* horses and sentenced to *two* years when he was *22*. Brought to jail by *two* men on April *22*, he was locked in cell block *2* on range *2*—in cell *22*.

Charmed Life

Patrolman Ambrose Dolney of the Detroit police escaped death by inches while arresting an escaped convict in March 1929. The man opened fire from just 4½ feet away, leaving 10 bullet holes in Dolney's coat. But he was unscratched.

DR.
J.I. GUILLOTIN
(1738-1814)
WHO INTRODUCED THE GUILLOTINE TO FRANCE.
WAS A PREMATURE BABY
-- BORN ON MAY 28, 1738,
WHEN HIS MOTHER BECAME REVOLTED AT
THE SIGHT OF AN EXECUTION

An Eye for an Eye

Many crimes involve an element of vengeance.

If you've been wronged in some way you can either take the matter to the courts or take the law into your own hands. Consider these chance encounters:

* A Brooklyn man shot four people who laughed at him when he mistook a jukebox for a cigarette machine.
* A library in East Orange, New Jersey, had 19 people hauled out of bed and booked in the middle of the night for keeping books overdue for a year.
* Canadian singer Cal Cavendish dropped 100 pounds of pig manure on the city of Calgary from a plane after losing his pilot's license.
* A rock promoter forgot to remove the brown M and M's from the two pounds of candy that the rock group Van Halen had asked for. So the group destroyed the dressing room and stage.
* A Venezuelan farmer hacked his pet goat to death because it ate $1,600 of his money.
* Eugene Schneider cut his $80,000 New Jersey house in half with a chain saw when his wife divorced him after 33 years.
* The audience started throwing bottles at a folksinger in Bangkok who was three hours late for a concert. So he opened fire on them, wounding three.

Good Insurance

M. W. Powers, police chief of Chattanooga, Tennessee, carried his hospital insurance in his breast pocket. And it saved his life by deflecting a bullet which had been fired at him.

Nice Shooting, Partner

Det. Sgt. Fred Tapscot of the Chicago police shot directly into the barrel of a convict's gun as the man was firing at his partner in 1922. The convict's gun jammed and the policeman's life was saved.

Must Be a Better Way

A Rochester, New York, policeman tried to stop a man from killing himself with a knife—by shooting him.

Fools of Fate

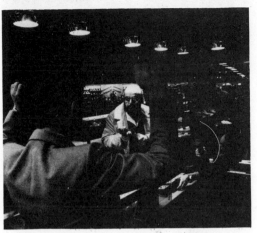

Some people just aren't cut out for the criminal life:
* A man held up an Ontario milk store disguised in a pair of ladies' panties—but stuck his face through one of the leg holes to demand money.
* Two teenage-girl muggers were arrested after they demanded payment in personal checks made out to one of them.
* Pierre Morvan was arrested because his money "smelled funny" when he tried to spend it in a store. He'd robbed an 80-year-old man who kept his cash in his socks!
* A man robbed a bakery in Elmont, New York, then put his sawed-off shotgun in his belt. As he drove off, he shot himself in the testicles.
* A 350-pound man held up a Long Island jewelry store, slipped as he tried to leave and couldn't get up before the police arrived.
* Two Chicago thieves left behind a camera when they burgled a house. The owner had the film developed—and it contained a photo one thief had taken of the other.

Cruel Fate

Los Angeles police shot and killed Donald Lee Oughton, who they thought was reaching for a weapon. But Oughton, who had a speech impediment, was trying to get out a card explaining his condition....

For the Defense

"When in doubt, tell the truth," said Mark Twain. But some truths sound more unlikely than the wildest fiction:

* The murderer of San Francisco Mayor George Moscone pleaded guilty by reason of insanity. His lawyer said his client had stuffed himself on Twinkies cupcakes and induced temporary insanity.
* A Miami lawyer said his client's planets were in opposition on astrological charts, causing him to knife a man, rape a woman and leave the house with a brassiere on his head.
* A London, England, man contested a divorce saying that the grunts coming from his bedroom were he and his girl friend playing snooker and exclaiming over difficult shots. The wife won—after testifying that the bedroom lights were out during the game....

The Law Is an Ass

Some places have some weird and wonderful laws. You'd probably get away with most of these things anywhere else, but:

* You can't open a massage parlor in Horney-town, North Carolina.
* You can't kiss or hug in public in Swedes-boro, Massachusetts. You also can't deface a statue without a permit or fire a cannon into Lake Narraticon.
* You can't keep a crocodile more than 12 inches long in Denver.
* You can't milk anyone else's cow in Texas.
* Animals can't have sex in public in Stanfield, Oregon.
* You can't make faces while wrestling in Los Angeles.
* You can't walk backwards while eating peanuts at a concert in Greene, New York.
* You can't walk, stand, lie or sit on the grass in public parks in Newquay, Cornwall.
* You can't feed coffee to a baby in a restaurant in Lynn, Massachusetts.
* Barbers can't eat onions from 7 A.M. to 7 P.M. in Waterloo, Nebraska.

Blind Justice

* Sir John Silvester (1745-1822), presiding judge of a London criminal court, had his pocket picked by a thief while he was finding him not guilty of picking pockets. The thief later confessed and returned the judge's watch.
* Harry Snyder was named head of a committee to screen films and view cabaret acts in Clarkestown, New York, to check for obscenity. Mr. Snyder is blind.

Not So Frank

Accused of making a patronage deal, former Philadelphia mayor Frank Rizzo took a lie-detector test to prove his innocence. He failed!

Head for Business

Highwayman Dominique Allier, of Le Velay, France, learned that the mayor was offering $600 for his head. So he visited the mayor's office, collected the ransom and escaped safely—with the money and his head.

Extensive Damages: Civil Suits

Chances are these crazy things will never happen to you. But they happened to some poor unfortunates—and they sued:

* Betty Penrole sued God for $100,000 after she and her Phoenix home were struck by lightning.
* Lady Portman, of London, England, sued her gardener for spanking her after he caught her "mucking about with my plants."
* Agnes Matlock of Long Island sued three fire authorities for arguing over jurisdiction while her home burned to the ground.
* A woman sued Coca-Cola for $15,000 after finding a hairpin in her bottle. She won.
* Gloria Sykes got $50,000 when she claimed a cable-car accident had made her a nymphomaniac.
* Janet C. Elliott sued Clairol for $5 million because, she said, Loving Care gave her abdominal pains, turned her urine black and damaged her vascular and nervous systems.
* Betty Hahn's husband was so embarrassed when she was praised on Ralph Edward's *This is Your Life* as self-sacrificing that he sued the networks for making a fool of him.
* An Oklahoma writer launched a suit for $2 million against a tavern owner and Coors brewery, claiming that their product pickled his brain and made it impossible for him to write.
* Knox County was sued for $25,000 by a woman who claimed a wall-mounted toilet fell to the floor with her on it, and that she now has to look for floor-mounted public conveniences.
* A court in Varese, Italy, ordered a male motorist to pay a 43-year-old prostitute $4,800 after a car accident that "reduced her professional capacity" by 15 percent.

Confessions: They Got It Wrong

The criminal has confessed. It's an open and shut case—until fate intervenes:
* Louise Butler confessed in gory detail that she'd murdered her niece, Topsy. But Butler was dead when Topsy turned up alive and well—the confession had been a figment of a diseased imagination!
* John Johnson confessed to the murder of Little Annie Lemberger in Madison, Wisconsin, in 1911. He'd served 10 years of a life sentence before it was discovered he had had nothing to do with Annie's death.
* A London hotdog salesman named Leonard Davey pleaded guilty to possession of marijuana—because he didn't know what sort of seeds he was carrying. He was on his way to Wormwood Scrubs Prison when it was discovered he'd been carrying grass seed!
* Jean Deshayes gave a detailed confession of a murder to French police in 1948. He later tried to retract it but the magistrate decided the story was so detailed and convincing it must be true. He'd been in jail for four years when a conversation overheard in a cafe led to the arrest of the real murderers!
* Louis De More had just arrived in Chicago when he overheard a policeman describing an armed robber who'd held up a tram and mortally wounded another policeman who was chasing him. He jokingly remarked that he fitted the description perfectly. The only witness identified him as the robber and the dying policeman nodded his head when he looked at De More. He pleaded guilty to get a light sentence and was jailed for life. But 10 days later the real killer was arrested for another crime and the murder gun was found on him!

THE ⑥TH VISCOUNT OF STRATHALLAN
1767-1851
on the Isle of Mull, Scotland
SWORE HE WOULD GIVE HIS GOOD RIGHT ARM TO WIN A LAWSUIT

HE WON THE CASE — AND A MONTH LATER A FLYWHEEL IN A FACTORY HE WAS INSPECTING *CUT OFF HIS RIGHT ARM BELOW THE ELBOW*

Drive on, James

In any assassination attempt on a VIP, odds are that a car is involved. More than 80 percent of assassinations and kidnappings are auto-related. In recent years, for example, President Ronald Reagan was about to get into a limousine when he was shot, pop star John Lennon had just got out of a taxi when he was gunned down, and Pope John Paul II was hit while riding in an open jeep in St. Peter's Square in Rome.

"As the world gets worse, my business gets better," says Tony Scotti, who offers a $1,650 course that trains bodyguards and chauffeurs how to act in an emergency. And not one of his 6,000 graduates in the U.S. and 33 foreign countries has been injured—though they've been involved in at least 44 such incidents worldwide.

Finders, Keepers

Two men were driving along in their car in Philadelphia in 1981 when they saw the back doors of a security van open and a trolley with $1 million on it roll into the street! They scooped up the money and fled!

Apparently the bumps in the road had triggered the automatic locks on the van doors!

The Big Steal

A 100-foot steel bridge spanning Piney Creek, Alabama, was stolen by two men, who sold it to a scrap yard for $149. The theft wasn't reported for two weeks.

Stop, Thief!

When James Watson's car was stolen in New York City, he set out to hitchhike home to Yonkers. The first car he flagged was his own. It did not stop.

Caught Red-handed

A California State University student was arrested in 1981 by customs officials when her clothing began to drip. She'd hidden three dozen slimy Giant African snails in a pair of boots, her purse and jacket lining. The snails are a delicacy in Nigeria but considered a pest in the U.S.

Computer Crime: The Microchips Are Down

Push the right button and, if Lady Luck is smiling, you could be an instant millionaire! The game of cracking computer codes, of proving yourself superior to a machine, has turned into crime on a grand scale.

A study by the Stanford Research Institute in the U.S. revealed that some 50,000 major crimes were committed with modern technology in the five-year period ending in 1972. Since only one in four people who commit such crimes is ever caught, the actual figure was probably nearer 200,000. And since then, it has soared even higher.

∗ An employee at a New York stockbrokerage programmed the computer to siphon off funds from a company account into his and his wife's accounts. In eight years he stole nearly $250,000. His employer found him out, but then together they milked funds from a rival company's computer, forcing the company into bankruptcy.

∗ Manfred Stein was accused of transferring stocks to his own possession, using a bedside telephone and teleprinter to talk to the computer. He told the computer to erase all traces of the transactions—but the message was intercepted by telephone engineers who by chance were making a routine check of his phone line.

∗ A German corporation paid $200,000 in 1973 to retrieve 22 reels of customer records and marketing data, stolen by one of its former computer operators.

∗ A ransom of nearly $1 million was paid for software snatched in Japan and ransoms totaling nearly $320,000 were paid in two similar cases in the U.S.

∗ The System Hackers, a gang of youths in the Los Angeles area, is being sought by police for committing more than $1 million worth of computer crime.

∗ Hero of the young computer criminals is John Thomas Draper, who discovered that a toy whistle vibrated at exactly the right frequency would unlock the vast network of the world's computerized phone system. Nicknamed Captain Crunch, he was jailed for using his computer to break into the U.S. military's secret phone system.

∗ Two pupils, one 16 and the other 17, at Palo Alto High School in Los Angeles's Silicon Valley, worked their way into the school district's computer. They weren't caught—they confessed voluntarily.

Taking the Credit

Scratch a credit card and you'll find an opportunity to commit crime. What are the chances, for example, of getting a card for the following person?

NAME: J. Christ
ADDRESS: 1 St. Peter's Gate, Heavenly Acres, U.K.
OCCUPATION: Superstar

Author Robert Farr had that card made up as a joke. But he accidentally presented it along with a check at a supermarket. The clerk accepted the card without comment—and cashed the check.

∗ A Toronto man applied for credit cards under five different names, went on a $22,000 spree between 1978 and 1980—then filed for bankruptcy when the firms involved asked for their money. None of the purchases or money was recovered.

∗ British bank and insurance executives estimate 2,000 checkbook and bank-card swindles take place each month in Britain.

∗ In 1973 one New York bank cashed $1,270,000 worth of checks written by thieves. Most were presented with stolen identity cards.

THE **IRONY OF FATE**
JEAN DE LA BALUE - INVENTOR OF THE "**IRON CAGE**" - A TORTURE CHAMBER IN WHICH THE PRISONER COULD NEITHER STAND UP NOR STRETCH OUT- **WAS THE FIRST MAN IMPRISONED IN IT!** HE SPENT 11 YEARS IN HIS OWN CAGE
Loches, France

Stamp of Success

* Jean de Sperati was such an expert stamp forger that the British Philatelic Association had to buy him out in 1954. During his career he turned out 500 varieties of stamps that were so good he was called into court several times to authenticate them.

* Raul de Thuin of Merida, Mexico, spent 30 years forging thousands of stamps that fooled the leading authorities. A block of six Canal Zone stamps was offered in 1966 for $1,900 and another "Mexican collection" was up for sale for $80,000 in 1962. He was never jailed.

* The greatest art forger of all time was the Dutchman Van Meegeren (1889-1947) whose work baffled the experts for years. He was unmasked only by chance after World War II. He'd sold a painting to Nazi Hermann Goering and was accused of collaborating with the enemy. To save his life, he had to admit he'd amassed a fortune by painting "Old Masters."

Monkey Business

A monkey was trained by French thieves as a housebreaker. It was finally caught when it left a fingerprint on a windowsill.

Hoaxes: It's a Swindle

All great hoaxes have one thing in common—a slice of luck. As Abe Lincoln said, "You can fool all of the people some of the time and some of the people all of the time, but you cannot fool all of the people all of the time." The great con man picks his target with care, takes a chance and in many cases a fool and his money are soon parted!

* Count Victor Lustig, an Austrian working for the French Department of Works, convinced businessmen in 1925 that the Eiffel Tower was to be sold for scrap metal and asked for tenders. He accepted the bid of Andre Poisson and then asked for a bribe. Poisson paid up—and Lustig disappeared.

* The Eiffel Tower was "sold" again in 1947—by a vegetable peddler to a Dutch syndicate. He actually collected a down payment of $150,000. Then he collected five years in jail.

* Scot Arthur Ferguson made a fine living convincing the gullible to buy parts of London—including Big Ben, Buckingham Palace and Nelson's Column. He went to the U.S. in 1925 and "rented" the White House to a Texan, but his luck ran out when he "sold" the Statue of Liberty to an Australian, whose banker advised him to check it out. After five years in jail, Ferguson retired from the con game and lived the rest of his life in luxury.

* In 1824 John Lozier and John DeVoe convinced hundreds of people on Manhattan Island that they should saw the island off from the mainland and turn it around. Why? The Battery area had become too heavy and was going to sink. They recruited tradesmen and carpenters and there was a march to the northernmost end of the island—but neither man showed up. It was a huge practical joke!

Twist of Luck

A safecracker who attempted to crack a vault by trying all 100 million combinations would have to work at it for 83 years.

The Thief of Time

The biggest thief of all? He or she may be staring back at you when you look in the mirror, according to Robert Half, a New York expert on personnel management. Half says current studies show that the U.S. is losing $98 billion a year through employee laziness and dubious illnesses. Each employee steals an average of 4 hours and 5 minutes a week, compared with 3 hours and 30 minutes in 1970.

Workers:

* Go slow to make jobs last;
* Moonlight;
* Come in late, leave early and take long lunches;
* Make personal phone calls at work;
* Say the boss does it, so why shouldn't we;
* Are just bone idle.

How to fight it? Hire employees only when you need them, make work rules at the start and stick to them, and find the worst offenders—and get rid of them first.

All Wrapped Up

French bank robber Raymond Burles collected $6,400 at gunpoint. But he was caught when he placed the money and his pistol in a small satchel and zipped it shut!

Carnegie Haul

Cassie Chadwick (born Betsy Bigley, daughter of a Canadian farmer) convinced people she was the illegitimate daughter of the great industrialist Andrew Carnegie, and was able to borrow millions of dollars from bankers. In her best years, it's estimated she raised—and spent—$1 million a year.

John the Ripper

English shoesmith John Bagford (1650–1716) spent all his spare time in libraries ripping illustrations and title pages from rare books. He filled 64 scrapbooks with them and the collection was so important that it was acquired by the British Museum.

Tricky!

* A bank robber in Queens, New York, asked the teller for his money in 10s, 20s and 30s!

* Two thieves planned in 1978 to "borrow" an empty truck and load up with cigarettes from a warehouse. Instead they picked on a truck already loaded—with 4½ tons of fire hydrants and two tons of toilet-training manuals called "I'm a Big Boy/Girl Now."

* A Portland, Oregon, bank robber pushed through the grille a note saying, "Put all the money in a paper bag." The cashier wrote on the bottom of the note, "I don't have a paper bag." The robber fled.

* Leo Koretz, a Chicago stockbroker, persuaded some of the city's wealthiest families to invest in phoney mahogany forests and oil in Panama. The scheme was launched in 1916, but it was not until 1923 that any investor thought of going to Panama to check on his investment. Koretz fled to Canada but was a diabetic and one of the few people trying a new drug called insulin. He was traced to a doctor and arrested in Vancouver.

* Ferdinand Waldo Demara had lived nine successful lives to the full by the time he was 30. Born in 1922, he played the roles of Trappist monk, deputy sheriff, psychologist, Christian Brother, doctor, soldier and sailor. He also became a ship's surgeon in the Canadian navy—his first job was to extract the captain's tooth—and a professor of biology.

* A U.S. prisoner who took an elementary course in engineering in jail invented a string of impressive qualifications for himself when he was freed—and landed a job designing the C5-A, the world's largest transport plane.

* A 25-year-old unemployed lumberjack and a 14-year-old runaway boy lived a life of luxury all over the U.S., with the elder posing as the son of then Canadian Prime Minister Lester B. Pearson, and the boy as his valet. They were given an escort for their Cadillac by the U.S. government as they robbed home after home, but were finally caught.

The Resurrection of John Stonehouse

Detectives recover a haul of stolen money.

The news made headlines round the world in November 1973. British M.P. John Stonehouse was missing off Miami Beach in Florida—presumed drowned. But Stonehouse was alive and well and about to start a new life in Australia. It was all part of a devious plot he'd worked out with his mistress, Sheila Buckley.

Stonehouse was desperate. A Board of Trade investigation was about to take place in England which would expose him as a liar and a cheat. His business would collapse and he would be ruined.

John Stonehouse's vanishing trick might have worked—but for an incredible piece of bad luck. The day after he arrived in Australia, he went to a bank in Melbourne to check that $24,000 Australian had been transferred from London to the account of Joseph Markham, one of the two new identities he had prepared for himself. He went into another bank and made a deposit, introducing himself as Donald Mildoon to bank teller Bryan King. King later spotted Mildoon in the first bank, and when Mildoon came to the second bank again, King became suspicious.

A check revealed that the first bank knew nobody by the name of Mildoon but that an Englishman named Markham had been drawing out large sums of cash. The police were told and Stonehouse was eventually arrested in August 1976 and jailed for seven years.

I Know That Name...!

N. W. Earl, a teller in a bank in Ottawa, Canada, spotted a forgery immediately when asked to cash a check for $54 in May 1961. The check had his own name and account on it.

Song of Freedom

Walter F. Robinson, an inmate of the State Prison Colony at Norfolk, Massachusetts, entered a radio song contest in 1950—and won a weekend trip for two to New York!

Calling All Thieves

Police investigating a burglary decided to try the number of a pocket paging device that had been stolen. It went off in the pocket of a man standing nearby—who was being questioned about a completely different crime!

In Their Debt

John Coffee built the jail in Dundalk, Ireland, in 1853, went bankrupt on the project and was the first inmate of his own jail.

How a "Sewer Rat" Was Trapped

The raid on the Nice branch of the Société-Générale was the biggest bank robbery in French history. Police believe that the thieves got away with $50 million—and most were never caught.

But Albert Spaggiari, the thief who had masterminded the whole thing, was caught—all because of a heavy downpour of rain!

The gang reached the vaults of the bank by digging a 24-foot tunnel from the nearest sewer. It took them two months. Finally, on Friday, July 20, 1976, they broke through the five-foot-thick walls into the vaults. They could have worked on undisturbed until early Monday—but for that shower of rain, which threatened to flood their escape route, a main storm drain. The gang made a hurried getaway Sunday, after going through 317 of the 4,000 deposit boxes in the vaults.

But they left behind thousands of dollars worth of equipment. Police traced Spaggiari through the shop where he'd bought the equipment—and by the Dom Miguel cigar butts that they found in the vaults.

However, Spaggiari escaped from a courtroom by jumping out of a window and reportedly went to South America. He became a folk hero in France and even wrote about the robbery, which was made into a film.

Bah, Humbug!

* Lee Garen of Fort Lauderdale, Florida, was mugged by 600 children while playing Santa at a shopping mall.
* Mary Olariu of Detroit, Michigan, drove 27 years without getting a ticket—then was given three traffic summonses in five minutes in April 1953.
* You can't buy a hat in Kentucky—unless your wife comes along to help you choose it.
* You can't fly a kite in Washington, D.C. If you do, you can be fined $5. And you could also be fined for carrying a woodcock there—alive or dead.
* You can't fight a duel with anyone in Uruguay—unless both of you are registered blood donors.
* You can't speak "English" in Illinois. The state legislature passed a bill in 1935 desig-

nating "the American language" the official tongue of the state.
* You could have kept a slave until recently in Kentucky, which has never accepted the 13th Amendment.

A Little Ingenuity

Fortune favors the bold. And these people chanced their hand—and won:
* A man who tried to rob a late-night store in Florida found there was only $50 in the till. So he made the clerks walk into the freezer and then waited on customers for three hours to boost his take. Among the customers he served were two local policemen!
* Patrolman Al Femenia was stymied when 15-year-old Sam Jones wouldn't surrender when trapped by Phoenix, Arizona, police. So he told the boy he was sending in vicious dogs, got the other policemen to bark—and the boy gave himself up.

Alone and Blue

Dani Elaine Trimarco of Tucson, Arizona, was being sought by the police for attempted fraud. She was found naked and shivering after tossing her blue clothing into an alley. Police had a description of the clothes and by stripping she'd hoped to blend in with the vegetation!

Cast-Iron Case

The Edgar Laing store, one of New York's first cast-iron buildings, was carefully dismantled in 1971 to be reerected elsewhere as a landmark—but thieves stole most of the iron sections.

PART SIX:

A Day in Your Life

Health, Happiness and your Daily Risks

You may think you are the master of your fate, but is it really true?

Come with us now as we explore the everyday things that make up our lives and look at the element of chance—a factor that can affect us 24 hours a day, shape our tomorrows and even dictate the way we live and die.

Still don't believe it?

Come then as we learn that the chances of living a long and healthy life are rooted in principles and beliefs that were once considered silly superstitions or outright heresies.

* Germ cultures were isolated because a scientist spilled a drug.
* Ether came to us because carnations wilted.
* Insulin was developed because a doctor was curious about a dog.
* Bandages were invented because a barber ran out of oil.

Even stranger are the stories of the chance discovery such as the one made by an exhausted Louis Pasteur, who brought us a cholera cure because he took a vacation. Or take the great surgeon Lister, a man who developed the process of sterilization because he hated dirty rooms.

What are the chances of living a long and healthy life? We learn women live longer than men and—more important—why. We learn the odds of beating lung cancer and about the strange case of Henrietta Lacks.

We meet Norm Cousins, a man who laughed himself to health. We meet others who cured disease by falling or reinjuring themselves.

For pet lovers, we find that dogs can keep you sane. For lovers in general, we find that kissing is healthy.

And we also find what kills us and why. One of the deadliest killers in America is the front-hall stairs; coffee tables injure 64,000 a year.

Birth, death, disease and accidents can all be intertwined with chance, fate or the unexplained. Read how it can happen in your daily life.

The World's heaviest twins, Billy and Benny McCrary of Hendersonville, North Carolina, professional wrestlers, had a combined weight of 1,474 pounds.

One Day in the Life of...

Everyday life is fraught with danger. From the moment we wake up in the morning to the time we switch out the light at night, we're taking chances in almost everything we do. Consider these American statistics.

* You wake up in the morning and reach out for the bedside light. That could be your first (and last) mistake: faulty wiring kills 500 people a year in the U.S.

* You head for the stairs. Careful now! Falls, particularly on stairs, kill 16,000 people a year.

* You pause at the door to the kitchen. One out of every 360 families suffered a door-related injury in 1977.

* Breakfast? Well, cancer-causing agents are suspected in coffee, tea, saccharin and red meat. Even peanut butter could be dangerous!

* You go into the living room. Mind that coffee table. They account for almost 64,000 injuries a year!

* You're setting out for work. By bicycle? Fatal accidents are more frequent on bicycles because they are unprotected. By car? One in four Americans has been involved in a car accident. By motorcycle? Worse still! One motorcyclist is killed every year for every 1,251 on the road. Your best bet is to take the bus. Odds on being killed on a long-distance bus are 21,000,000 to 1, on a local bus 231,000,000 to 1.

* A walk in the fresh air may do you good. But probably not if you live in the eastern U.S., where air pollution kills 20,000 a year.

* You made it to work. Congratulations! But take care—there were 2.2 million disabling accidents at work in 1978, according to the U.S. National Safety Council. Of these, 13,000 were fatal while 80,000 resulted in permanent impairment.

* Driving home? Nearly half of all fatal vehicle accidents take place between 4 P.M. and midnight, and over half of the deaths (54 percent) occur on Friday, Saturday and Sunday.

* A cigarette and a beer to celebrate getting home? Cigarettes cause 40 percent of all cancers and kill 15 percent of all Americans. Thirty percent of Americans who smoke die of cancer or heart disease due to smoking. A cigarette takes 10 minutes to smoke—and shortens your life expectancy by five minutes! And that beer may cause cirrhosis of the liver and may come in a cancer-causing bottle.

* And so to bed. Perhaps you should have stayed there in the first place. Or perhaps not: most people die in bed. And beds caused 77,581 injuries in 1977 while pillows caused another 445.

It Could Have Been Worse

Clean-air regulations saved 14,000 lives in the U.S. in 1978, says Prof. A. Myrick Freeman III. The saving on medical bills was $21.4 billion, compared with the $16 billion cost of the cleanups.

And according to the Massachusetts Institute of Technology Center for Policy Alternatives, better laws and regulations have saved:
* 40,000 to 60,000 working days, and 630 to 2,500 workers from lung cancer and asbestosis.
* 28,000 from auto-related injuries.
* 34,000 drug-related injuries to children.
* 44 percent of crib deaths.

More Accidental Facts

* Only 1 percent of all deaths in Ontario, Canada, occur in the home, but accidents in the home cost the province more than $292 million in lost wages, expense and inconvenience—about $50 for every man, woman and child.
* Poison has been declining as a cause of serious injury.
* Most falls occur between 10 A.M. and 3 P.M. They usually happen at the end of the week—and at the end of the year.
* In Canada, six people are blinded each day in accidents.
* Exploding car batteries caused more than 6,000 eye injuries in Canada and the U.S. in 1980.
* 80 people in the U.S. and 20 people in Britain die each day in accidents in the home.

Help

Everyone gets a flat tire on a highway sooner or later, but did you ever consider your chances of flagging down help?

A German magazine staged a test and found out that an attractive young woman finished in fourth place behind a 70-year-old gentleman. In second place was a nun and third went to a pregnant lady.

What's That?

Noise is all around us; there's not much hope of escaping it. And it's probably bad for our health.
* A University of Miami team exposed monkeys to the kind of noise we face every day—radios, traffic, noise in the office and factory—and measured its effects on their health. Their blood pressure went up 25 percent and their heart rate changed.
* Monkeys exposed to rock music at first jumped around, became annoyed and then tried to follow the beat and finally wound up staring sullenly into space.
* If we didn't have good insulation, our own heartbeat would deafen us. Fortunately the inner ear is bathed in fluids that absorb and deaden the noise.

Women Last

Men may be stronger physically, but that's about the only biological advantage nature has given them.

* Males are more vulnerable before they are born. It's estimated that 140 males are conceived for every 100 females but only 106 born for every 100 girls.

* 15 per 1,000 boys die in the first month of life, compared with 11 per 1,000 girls.

* Of 190 birth defects observed, more than 71 percent occur mainly in boys while only 25 percent are found chiefly in girls.

* Heart disease affects a significantly higher number of men than women, in every age group.

Three score Years and 10

The Bible gave man's life span as threescore years and 10. But it's only recently that the average life expectancy has soared to 70.
* Prehistoric man lived only 18 years on average and rarely survived to 40.
* The ancient Greeks averaged 20 years, while in ancient Rome it was 22 years.
* In the Middle Ages in England, you could expect to live 33 years, and the average life span was up to 36 by the mid 17th century.
* By the turn of this century, it was 47 in both England and North America.
* By the end of the 1930s it was up to 63.3, and by the end of the 1940s it had risen to 68, thanks to the discovery of antibiotics.
* The average life expectancy now is 70, with women living 7.8 years longer than men. Many killer diseases, like typhoid, malaria and TB, have been controlled or eliminated.

YAUPA

A NATIVE OF THE ISLAND OF FUTUNA, IN THE NEW HEBRIDES. DIED IN 1899 AT THE AGE OF **130** AFTER CONTRACTING MEASLES --*A CHILDREN'S DISEASE*

© King Features Syndicate, Inc., 1971 World rights reserved

Bill Malbert of Moldgreen, England at the age of 103 years in 1957.

It's a Graying World

Shigechiyo Izumi of Kagoshima, Japan, who was 115 years old in 1980, and reputedly the oldest man in the world.

Your chances of making 65 are excellent. Every day, 5,000 Americans celebrate their 65th birthday, making a net increase of people over 65 of 1,600 every day, 600,000 a year. By the year 2000, there will be 32 million people over 65.

And the whole structure of society will change between 2010 and 2030 as the baby-boom generation turns 65.

A Long and Happy Life

Your chances of living past 100 are much higher in certain areas of the world, and medical and nutrition experts say diet has a lot to do with it. For example:

∗ In Hunza, Pakistan, the oldest citizen is 110, the second oldest 105. People there have a simple, low-calorie, low-fat diet.
∗ There are more people over 100 in the Caucasus Mountains of Russia than anywhere else in the world. They eat simply, have a varied diet and a wide change in yearly climate. They also celebrate feasts and toss back five brandies at a time.
∗ Nine out of every 810 people in Vilcabamba, Ecuador, are more than 100. One out of every six is over 60. They eat a low-calorie diet, smoke and drink moderately and most are of European descent.
∗ Oh yes, in North America, three out of every 100,000 live past 100.

Medical and Personal Problems

Sooner or later, everyone has a medical or personal problem. But the chances you'll have an unprecedented case are slim indeed. Most fall into categories and scientists can now predict your chances of contracting a given disease or undergoing a certain type of operation.

The single most common type of operation listed by the British Department of Health is an episiotomy, or assisted birth. It's followed by operations for a tipped uterus and a hernia, and then come bladder surgery, appendectomy, pregnancy termination, tonsillectomy and removal of the gallbladder.

The most frequent reason for going into a hospital, however, is normal delivery of a baby, outnumbering surgery by almost 2 to 1.

The chances of a 35-year-old woman dying of heart disease are 3,270 to 1, but that shortens drastically after age 65 to only 48 to 1.

Women, however, have many more personal problems than men and the chances are high they'll encounter a crisis at some point. A London newspaper monitored personal problems in its lovelorn column and found matrimonial woes were the most common, closely followed by loneliness, jealousy, teenage problems, in-laws, homosexuality, depression, physical appearance, agoraphobia, unmarried mothers and employment.

MAY LORETTA WILLET of Ottawa, Ill., WAS BORN MAY 29, 1897--
HER GREAT-GRANDDAUGHTER LEIGH ANN ZINANNI WAS BORN MAY 29, 1963 --HER GREAT-GRANDDAUGHTER KRISTY LIN ZINANNI WAS BORN MAY 29, 1965 --HER GREAT-GRANDDAUGHTER SHANNON GERRALD WAS BORN MAY 29, 1974

Standing Room Only

Unless fate intervenes in the shape of a nuclear holocaust, the world population will continue growing. It has now reached a record 4.5 billion and is climbing at a rate of 1.73 percent a year, according to a new United Nations study. It will probably take 130 years to level off, at about 10.5 billion—with 2½ times as many people on earth as there are now!

* The industrialized nations of Europe and North America are growing far more slowly than the underdeveloped countries—0.6 percent compared with 2.1 percent—mainly due to increased prosperity, education and birth control.

* Europe will achieve a stable population in about 50 years, North America in about 80 years, East Asia in 110 years, the Soviet Union, Latin America and South Asia in 120 years, and Africa in 130 years.

* In the year 2000, the population of North America will be about 286 million, compared with 254 million now; in Europe it will be 511 million, compared with 486 million.

The Way of Death

How will you die? Well, here are the odds, according to the U.S. Surgeon General's Office:
* 50 percent of Americans die of "unhealthy behavior," e.g., killing each other.
* 20 percent by nature and the environment.
* 20 percent by biological causes.
* 10 percent due to inadequate health care.

The War to End All

In an all-out nuclear war between the United States and the Soviet Union, your chances of surviving in either country are virtually nil, particularly if you live in a major city. Here's why:
* By 1985, the Soviet Union will have about 6,000 nuclear warheads which they can aim at the United States. The U.S. has some 2,000 cities whose population is over 10,000. And 300 of these cities contain 60 percent of the population.
* More than 160 million Americans and 100 million Soviets would die in a war between the two, the then U.S. Defense Secretary Harold Brown told Congress in January 1981.
* Millions more would die later from fallout, polluted water and air, and firestorms, which would extend the area of destruction by a one-megaton bomb from 80 square kilometers to 400 square kilometers.
* Most hospitals are downtown and would be destroyed. There would likely be only one doctor left for every 2,000 people.

"Truly the living would envy the dead," says Dr. Bernard Lown of International Physicians for the Prevention of Nuclear War.

Blue Mondays

It's official. Mondays are dangerous to your health.

More men die suddenly of heart disease on a Monday than on any other day of the week, according to researchers at the University of Manitoba, Canada. And fewest men died of heart attacks on Fridays. A survey of nearly 4,000 men found that 38 had died of heart attacks on Monday while only 15 died on Friday.

Luigi II, duke of Savoia (1402–1465) slept 22 hours a day for the last 20 years of his life, only rising briefly for his 3 meals a day. Believe It or Not!

Edward Moore (1712–57), the English dramatist, although apparently in perfect health, sent his own obituary to the newspapers as a jest, giving the next day as his date of death. Moore suddenly became ill, and indeed died the next day. Believe It or Not!

Good Habits

Want to live a long and healthy life? Here are some tips on how to improve the odds, courtesy of the U.S. National Center for Health Statistics:
* Sleep seven to eight hours a night (68 percent of people who live long lives do).
* Eat breakfast (60 percent do).
* Snack (40 percent do).
* Exercise (again, 40 percent do).
* Don't smoke (50 percent never did, over 75 percent don't anymore).
* Don't drink (21.5 percent of men and 34.2 percent of women don't—but 43 percent of men and 18.5 percent of women drink more than five drinks at one sitting!).

Those who obeyed five out of these six rules were healthier and lived longer than those who followed four or fewer.

Too Many Mouths

If you live in Upper Volta, your hopes of surviving are grim. This tiny African country is one of the poorest nations in the world.
* You would have a 50/50 chance of dying before the age of 5 of either malnutrition or childhood diseases.
* If you survived, you would be poor, hungry and probably illiterate.
* Although attempts have been made to boost efficiency, the country's gross national product has dropped for the past 10 years.
* Your income for the year would be $201, half that of most developing countries.

Under the Weather

Those logical Germans have figured it out: it's the weather that makes you sick. Weather forecasts in West Germany now include the weather most likely to cause allergies, epilepsy and stomach pains, as well as heart trouble and colic.

Dr. Inge Goldstein of Columbia University has found that there are "asthma epidemic days" in New York City and New Orleans.

Dr. Anita Baker-Blocker of Illinois has found that heart attacks go down by 20 percent after a winter rainstorm.

Suicide: To Be or Not to Be

Suicide attempts are often a cry for help. Many people expected to be found and stopped—but chance stepped in and their cries were not heard. For every suicide there are at least 15 suicide attempts.

* The highest suicide rate in the world is in Denmark, where one in 4,000 people commits suicide. That's 26 per 100,000 inhabitants.

* The U.S. rate is 12, while in Britain it is 8. But in San Francisco the rate is an amazing 2,561 per 100,000.

* Teenage suicide is the biggest killer of youth in North America, after accidents. The number of suicides has increased 50 percent in the last six years in Canada alone. Three times as many girls try it, but boys succeed four times as often.

* Children of divorced or separated parents, of alcoholics, and of parents of whom one or both were suicides are the likely candidates for attempted suicide.

* Suicides who fail the first time are likely to "try hard" the next time—and succeed.

* Most girls die of drug overdoses while most boys hang or shoot themselves.

Fate Was a Jump Ahead

Alone and broke at Christmas, artist John Helms flung himself off the 86th floor of the Empire State Building in New York in 1977. He woke up half an hour later to find himself on a 2½-foot ledge on the 85th floor—he'd been blown back onto the building by strong winds! He knocked on the window of the offices of a TV station and crawled in to safety.

And there was a happy ending. Hundreds of families called to offer him a home for the holidays.

Appearance: You're Looking Good

Psychological studies confirm that your chances of success in life are far better if you're attractive. Among their conclusions:

* Job applications by good-looking people are picked over those from more homely people, even when qualifications are exactly the same.

* Teachers treat good-looking students better and they are more popular with their fellow students.

* More people will stop and help an attractive person who is in trouble than a plain person.

Feeling Good

Attractive people actually do tend to perform better. They have self-confidence and a good image of themselves.

And your personality affects the way people see you. A dynamic, outgoing person is often thought appealing, regardless of the way he looks.

Gentlemen Prefer Blondes

If you happened to be born a blonde, you're lucky. Blondes really do seem to have more fun. For example, a 1979 study tested the reactions of a group of men from various backgrounds and of different ages to blondes and brunettes. The response to the blondes was overwhelmingly sexual while the dark-haired women brought a variety of reactions.

Changing the color of your hair is very popular in the U.S. It's estimated that 45 percent of women there used hair colorants in 1975 while in Britain the number was 23 percent.

Squeaky Clean

Who has the cleanest hair of all? Canadians— over 95 percent of the population uses shampoo! For the U.S., Sweden and Japan, the figure is at least 90 percent. In Britain, it's 84 percent.

The Height of Success

The average man is 5 feet 9 inches tall, the average woman, 5 feet 4 inches. If you're taller than that, the gods have smiled on you. Sheer height can mean success.

One survey at the University of Pittsburgh revealed that the "bonus" for being 6 feet 2 rather than 5 feet 11 was a 12 percent higher starting salary among male graduates of one class.

And tallness gives you a head start in other areas too. You are more likely to be welcomed into clubs, win the best jobs and achieve high office!

Those Bright Blue Eyes

Dark-haired, dark-eyed people are more emotional and have quicker reactions than blue-eyed blonds, possibly because of a lack of pigment interfering with visual cues to the brain. But blue-eyed people have more stamina, are more sentimental and more often get bogged down by routine, according to a Chattanooga Institute report. They are also moodier and hold grudges longer.

The Bald Facts

If you're bald, the odds are you can't do much about it. In men, 95 percent of all baldness is natural—the other 5 percent is caused by such things as chemotherapy, glandular disorders, X-rays or chemical abuse.

Everyone gets bald. It's just that some are unluckier than others—and it shows. Men and women begin to lose their hair in their early 30s; by 50 half the hair can be gone; by 65, two thirds.

Baldness is hereditary. It may skip a generation and show up in the next. But you may be better off without hair anyway; doctors say it is filthy and spreads diseases!

You can always try to disguise the bald fact—and hope that nobody notices:

* Wigs. Louis XIV had 40 wigmakers and nearly everyone in the 18th century wore a wig. So did the Egyptians—though they first shaved their heads to get rid of lice.

Wigs have made a comeback in the U.S. thanks to such beauties as Jackie Kennedy and Farrah Fawcett. People despaired of matching their thick and luxuriant locks so they turned to wigs. The cost? $500 and up for a good one.

* Toupees. They cost $50 or more. The better the quality of the hair, the higher the price.

* Hair-weaving. The first one to know about it is your hairdresser. And so does everyone else before long, because the rug loosens as your hair grows. It has to be retightened frequently and is therefore expensive.

* Hair transplants. A little off the sides and the back, and onto the top. Hair is uprooted with its skin and transplanted onto the top of the head. A little goes a long way, since the tufts at the ears never stop growing. A careful surgeon and lots of money generally assure you of success.

Pie in the Sky

Colonel Morton's Kentucky-style lemon cream pie contained no lemon, no cream and no eggs, according to the Federal Trade Commission, the U.S. consumer watchdog.

And there was no Colonel Morton either.

Shedding Those Pounds

Thinking of dieting? Millions try it each year, shed a few pounds—and put it all back on again when they return to their normal eating habits. Your chances of keeping off those lost pounds are slim. Only 10 percent of dieters do, according to obesity specialist Dr. Theodore van Itallie of St. Luke's Medical Center in New York.

In most cases the weight loss in the first weeks of a diet is water, not fat. And the only long-term answer seems to be to find a nutritionally balanced way of eating that you can follow for the rest of your life.

Weighty Problems

Few things are as subject to the whims of fashion as weight. "Let me have men about me that are fat," cried Shakespeare's Julius Caesar, who didn't trust hungry-looking people, shortly before he was assassinated. Fat was certainly beautiful in the eyes of the 17th-century painters Titian and Rubens. The pendulum swung the other way with a vengeance in the 1960s and 1970s; the beautiful people were all slender.

Nowadays, the average man weighs in at about 170 pounds, the average woman at about 140 pounds. And the more you earn and the older you get, the more you tend to weigh.

But some doctors are now saying that a little fat is natural and that body weight depends on many variables. Studies concerned with heart disease have revealed that the average American man aged 40 to 59 is 17 percent above the desirable weight for men of medium frame. Yet doctors are now reluctant to link overweight with coronaries and early death.

One thing is for sure. Overweight people do tend to have high blood-pressure.

> **A FAT PERSON** WILL GAIN BETWEEN 5 AND 10 POUNDS A YEAR IF HE EATS ONLY ONE PER CENT MORE THAN HE BURNS UP

The Fat's in the Fire

What are our chances of becoming fat? Considerable, judging by the following:

* American executives eat too many carbohydrates and animal fats—bacon, eggs, doughnuts, potatoes, pies and bread. Obesity in some form is common and heart disease is a major killer.

* The Irish laborer works hard, but drinks an average nine pints of beer or ale daily, plus four cups of tea with milk and sugar. He rarely eats fruit or raw vegetables, except cabbage. Thus, he frequently dies of hardening of the arteries.

* Chances are three in 10 you're overweight if you're American, two in 10 if you're European.

* Three in 10 fat adults were fat children.

* Children often get fat because mothers stop breast-feeding and substitute milk and cereals, or use food as a reward in training.

* Despite obesity statistics, each person in the world on average consumes 2,751 calories a day. Yet he needs 2,900.

Brains and Brawn

Fat children tend to have higher IQs than skinny or average-weight kids, according to a study of 20,000 children in the U.S. conducted between 1959 and the early 1970s. At the ages of 4 and 5, the heaviest had IQs averaging 10 points higher than those of the lean children.

Figuring It Out

Working out complex mathematical sums in your head takes a peculiar kind of mind. But sometimes nature turns up idiot geniuses—people of subnormal intelligence who can do such sums in a flash. Here are some of them:

* Charlie and George, identical twins born prematurely in 1939, had IQs of between 60 and 70. Yet when tested at Letchworth Village, a state institution in Thiells, New York, in 1963, both could tell within seconds how much time had passed since anyone's last birthday. They could rattle off the years in which a given date would fall on a Sunday and the day of the week for dates as far in the future as 7000 A.D. They could also go back in time for hundreds of years—despite knowing nothing of calendar changes. Yet they couldn't subtract 10 from 20!

* Jedediah Buxton, an English-born "idiot savant" of the 18th century, never learned math, yet could perform complex calculations mentally for 2½ months at a time.

* In the 19th century, 10-year-old Truman Henry Safford was asked to multiply 365 x 365 x 365 x 365 x 365 x 365. In not more than a minute he came up with the right answer: 133,491,850,208,566,925,016,-658,299,941,583,225!

Some idiots savants excel at other things:
* J. H. Pullen, a 19th-century savant known as the Genius of Earlswood Asylum, designed and built a 10-foot model of a ship with more than a million parts.
* A 22-year-old epileptic studied in 1898 could repeat fluently lengthy passages read to him in three languages he didn't understand.
* Harriet, a Boston hospital kitchen worker studied in 1970, could play Happy Birthday in the styles of composers such as Mozart, Beethoven, Schubert, Debussy, Prokofiev, Verdi and others.

How do they do it? Well, incredible powers of concentration and memory may play a large part. Some researchers also believe that the idiot savant's special ability may be a compensation for poorly developed other faculties.

Tying and Untying the Knot

Marriage is a dicey business. Of all Americans over 50, only 6 percent of men and 4 percent of women have never married. But then a massive 38 percent of first marriages end in divorce.

Other facts about wedded bliss:

* American men are the youngest bridegrooms in the world, with an average age of 23.4. The average U.S. bride is aged 20.7. In Britain the average man marries at 25.5 and the average woman at 21.7.
* Spanish men wait the longest to get married—until they're 28.3 years old.
* Japanese brides are the oldest at 25.2 years.
* The vast majority (four out of five) of divorced Americans remarry, three out of 10 within three years.
* In 1978, 2.2 million Americans were living together without being married—an eightfold increase over the previous eight years.
* The divorce rate in the U.S. is the highest in the world, with 4.6 per 1,000 inhabitants in 1974. In Canada the rate per 1,000 inhabitants was 2, while in Britain it was 2.14.
* In Sweden there is one divorce for every 7 marriages.
* Catholic Italy, there's only one divorce ery 26 marriages.

Marriage: The Chinese Puzzle

Women over the age of 28 have little chance of marriage in Communist China. Why? Government laws on birth control require a girl to marry in her mid to late 20s. Many are condemned to a life of loneliness if they're over 30, or a desperate scramble to find a partner—any partner.

Oh, Sister

The Feminist Capital of the World? It's San Jose, California (population 628,000), which has a woman mayor, vice-mayor, town clerk, deputy city manager, labor director and seven women on its 12-member town council.

The Hole Truth

The city of Woonsocket, Rhode Island, voted to call the circular metal covers in its streets "personholes" instead of "manholes."

In the Know

More than half of Americans believe in angels, 39 percent believe in devils, 37 percent in precognition, 29 percent in astrology and 10 percent in ghosts and witches, according to a recent Gallup poll.

Till Death Us Do Part

Men have a much better chance of living longer if they're married rather than single, according to a 12-year Johns Hopkins University study published in 1981. Men between the ages of 55 and 74 who had lost their wives had a 60 percent higher chance of dying sooner than men in the same age group who were still married.

There was no significant change in the death rate of women whose husbands had died, the survey of 4,032 widowed people showed.

Unhappy Landing

Vera Czermak jumped out of her third-story window when she learned her husband had betrayed her—and survived. She landed on her husband, who was killed....

A Royal Wedding

Inside St. Pauls Cathedral during the wedding ceremony of The Prince of Wales and Lady Diana Spencer.

Close Encounters

Found the perfect partner? You're exceptionally lucky, at least according to *The Hite Report on Male Sexuality*. Of·the 7,239 men surveyed:

* 41 percent resented women.
* 54 percent felt guilty about their behavior towards women.

Annie Jane GIDDENS — Hohira, Ga.

WAS BORN IN THE *SAME HOUR* OF THE *SAME DAY* OF THE *SAME WEEK* OF THE *SAME MONTH* IN THE *SAME ROOM* OF THE *SAME HOUSE* AS *HER MOTHER WAS*!

—— AND SHE WAS ATTENDED BY THE *SAME DOCTOR*!

Close Encounters II

Some sexual statistics from other studies:
* Just over 50 percent of couples have sex from six to 15 times a month. One couple in 12 have sex 20 or more times a month.
* 57.6 percent of women say the frequency of intercourse in their marriage is about right while 38.1 percent wish they had sex more often.
* 29 percent of married women have had an affair. Of married men in the U.S., including those now divorced, one out of two has had at least one affair.
* Given a choice between their husband, a good book or other activities, 37 percent of women chose the book, 26 percent the husband. Men opted for sex (45 percent), with going to a sports game a close second (41 percent).

Water Baby

A 15-year-old girl gave birth to a baby boy nine months after swimming in a public pool in Sydney, Australia—although doctors said she was still a virgin. The courts ruled she had been impregnated by male sperm in the water!

Lasting Impression

When a black baby was born to a white couple in Munich, the husband demanded a divorce. But a gynecologist solved the mystery. The husband, he said, had had intercourse with a black prostitute, had picked up sperm left by a previous black client and transmitted it to his wife!

Births: It's a Boy!

Do sons run in families?

Yes, according to one study at Oxford University. Records showed that families with six or more sons and no daughters are twice as common as the laws of chance dictate.

Births and birthdays produce some incredible coincidences. For example:

* Emory Harrison of Johnson City, Tennessee, had a family of 13—all sons. Odds against this? Over 4,000 to 1.

* Seven generations of the Pitofsky family in New York have produced only sons. The 47th consecutive son was born in Scarsdale, New York, in 1959—a 136,000,000,000,000-to-1 chance!

* Mrs. Anna Andzurka of Cruszyna, Poland, had a baby on February 5, 1934. So did her daughter. And so did her granddaughter!

* Mrs. Margaret Connelly had a daughter on July 28, 1947. She had another daughter on July 28, 1948, and a son on July 28, 1949. Her own birthday? July 28!

* J. Ryan, Jr., of Elyria, Ohio, was born on January 26, 1942. His father, grandfather, great-grandfather and great-great-grandfather were also born on January 26!

MORTEN SOLVSBERG —Age 11 IS THE *FIRST SON* OF THE *FIRST SON* OF THE *FIRST SON* OF THE *FIRST SON* OF THE *FIRST SON* OF THE *FIRST SON* OF THE *FIRST SON* OF THE *FIRST SON* OF THE *FIRST SON* OF THE *FIRST SON* OF THE *FIRST SON* OF THE *FIRST SON* OF MORTEN SOLVSBERG WHO LIVED EARLY IN THE 17th CENTURY THEY ALL EITHER LIVE OR HAVE LIVED ON THE SAME FARM IN NES, HEDEMARKEN, NORWAY.

It's A Baby

All fathers like to brag, but there are some incredible stories about births. Call it chance, fate, luck or mystery, but we learn that:

∗ Louis Henri de Bourbon lost his left eye in a hunting accident in France and later became the father of five children in the early 1700s. Each child was born with a blind left eye.
∗ George McDaniels and his entire family— father, mother, sister, two brothers and an uncle—were all born on the same date: March 23.
∗ Ethile Hecock was born on May 18, 1929, in Ida Grove, Iowa—in the same town, at the same hour, on the same day of the same month with the same doctor and nurse in attendance as when her mother was born 24 years earlier.
∗ Edward Hanson of Eddington, Pennsylvania, was born on Easter Sunday, March 23, 1913. His birthday will never fall on Easter again during his lifetime.
∗ Mrs. Sandy Charles of Cuba, New Mexico, gave birth to twin daughters—84 miles apart. The first was born naturally in Cuba, but a cesarean was needed for the second, so Mrs. Charles was rushed to hospital in Albuquerque.

Personality Plus

Ill-tempered, edgy or angry? Happy, outgoing or romantic? Whatever your personality, the odds are you picked up most of it in the womb.

New medical research has shown that the personality is strongly affected by the level of a chemical called monoamine oxidae, a protein that helps control the brain's chemistry.

The protein comes from Mom before birth so it's possible that she's responsible for your good—or bad—temper.

Midwifery

If you live on the West Coast of the United States, the chances are one in five that you will have your baby delivered by a midwife.

The growth of the practice, established through the ages, is directly related in North America to the popularity of home births, where the newborn can quickly and easily become familiar with his home and parents. Midwifery has always been popular in England and is considered an extremely reliable and informed practice.

Also becoming popular is the new phenomenon of eating the placenta after the birth. Medical science reports it is becoming popular with white, middle-class, college-educated parents—even though the placenta tastes a bit like liver.

Among the Thais of Indochina a baby boy must decide his future vocation on his first birthday. Symbols of various occupations are put in front of him, and the object he first grasps will determine his life's work. Believe It or Not!

Being Born

How often have we heard the phrase "he's lucky to be alive"? People throughout history have been fascinated with birth, protecting the newborn and hastening the blessed event.

Not willing to trust entirely to chance, societies have come up with these customs to guarantee a safe birth:

* A woman smells her husband's sweaty work clothes in one African tribe to convince the unborn child that there will be a father to protect him.
* In another African society, a woman's vagina is packed with cow dung to show the unborn that the father is rich in cattle.
* Many tribes on the Dark Continent force a pregnant woman to confess all her sins so they won't affect the child.
* In biblical times, if a wife was barren the husband often took her servant. The servant was used as a surrogate mother, delivering the baby while lying in the wife's lap.
* The Near East believes in hurrying a delivery. A woman used to be tied to a stake while her husband rode past on a horse at full speed to scare her into labor.

* The entire courts of Louis XIV and XV had to be present during delivery to verify the offspring was indeed royal and not a commoner smuggled in from a village mother.
* In New Guinea, a husband must simulate his wife's labor pains while she prepares to deliver the child.

Afterbirth

Just as they take few chances with the unborn child, societies are most particular about the placental tissue which remains:

* In one West African tribe, the placenta is buried in a secret place because it is believed to contain part of the spirit of the child and thus has to be protected from evil. In another tribe the placenta is taken to a witch doctor who tells the fortune of the newborn.
* A tribe in the Philippines forces the father to bury the placenta to create a stronger bond between himself and the newborn.
* Modern hospitals, however, have a simpler view. The placenta is usually sold to drug or cosmetic companies because of its high estrogen content.

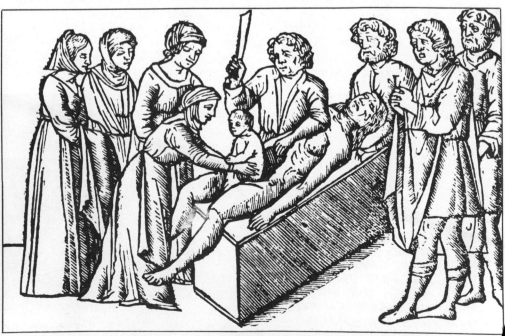

Early illustration of a caesarian section, showing the birth of Julius Caesar.

Multiple Births

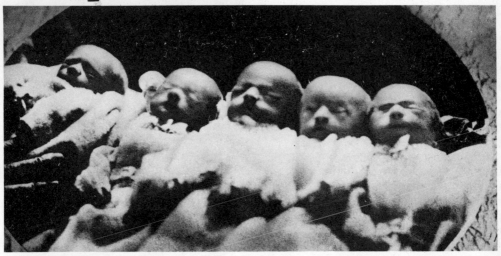

Quintuplets born February 15, 1880 in Nova Scotia. Three died the day of their birth, one a day later, and one three days later. Their poverty-stricken farmer parents turned down an offer of $20,000 from P.T. Barnum, who wanted to exhibit the bodies.

With new fertility drugs, multiple births are becoming more and more common. Years ago, however, they were decided by chance.

* A German woman in the 1600s gave birth to six children.

* Seven were born to two African women, one case in 1907, the other in 1920.

* A Mexican mother bore eight children in 1921 and the feat was repeated in China in 1934.

Unfortunately, the chances of survival of the tiny children were slim before modern obstetrics.

It was not until 1934 in a small Canadian town near North Bay that a case of quintuplet births was recorded where all children lived. Those were the famous Dionne quints and their incredible births stunned the medical world. In Great Britain, not even quadruplets managed to survive until the four St. Neots babies were born in 1935.

Now the Odds

Your chances of being a twin are one in 90; of being a triplet, one in 7,000; of being a quad, one in 500,000. One set of quints is born, on average, in every 40 million births.

'THE GEMINI TWINS BRIAN AND BRUCE PERRY WHO LIVE ON GEMINI DRIVE IN BASTROP, LA., WERE BORN ON JUNE 12th -THE MONTH THAT HAS THE ZODIAC SIGN GEMINI— AND GEMINI IS THE SIGN OF THE TWINS

German emperor Sigmund (1368–1437) was engaged to marry a count's daughter when he was only 3 days old. Believe It or Not!

Twins: A Whim of Chance

The fate of identical twins is decided by the 10th day of pregnancy. If two placentas form before day 10, the twins will be born identical. But if the placenta does not divide until the 13th day, the twins will be mirror-image, facing each other in the womb with opposite characteristics. After the fateful 13th day, the pair will be joined in the placenta and be born Siamese twins, forcing doctors to use surgery to separate them.

* The Van Arsdale twins played professional basketball for 12 years, mostly for different teams and with different coaches. Yet their National Basketball Association records were remarkably similar. Dick made 5,413 baskets, Tom made 5,505; Dick played 921 games, Tom 929; Dick averaged 16.4 points a game,

Tom 15.3; Dick had 3,807 rebounds, Tom 3,948.

* Twins John and Arthur Mowforth both had severe chest pains on May 22, 1975, and were rushed to hospitals in Bristol and Windsor, England. Each died of a heart attack shortly after arriving at hospital. Neither they nor their families knew what was happening to the other twin.

* An identical twin was in Los Angeles when her sister died in one of the two 747 jets that collided in the Canary Islands on March 27, 1977. She said her body "was enveloped with heat and intense pain." And she felt " black void" at the very moment her twin was killed.

Double Trouble

There are an estimated 100 million twins on earth. And they're a fascinating field for the study of the laws of probability.

The two types of twins are *identical*—who come from one fertilized egg and are always the same sex, blood type and look the same—and *fraternal*, who come from two eggs fertilized at the same time and have no more physical resemblance than any other children of the same parents. Fraternal twins account for two thirds of all twin births.

Here are some of the amazing coincidences involving identical twins:
* The "Jim" twins of the U.S. were brought up apart. Yet a 1979 study at the University of Minnesota discovered that both married girls named Linda, both divorced them and married girls named Betty. One called his first son James Alan, the other called his James Allen. Both named their dogs Toy. Both did well in math, liked woodworking, had headaches and took Valium.
* Twin sisters LaVelda and LaVona Rowe of Chicago married twins Alvin and Arthur Richmond. The couples live in the same house and the men say that if one is hurt, the other also feels the pain.
* Another set of twins both became fat in high school, but slimmed down later. Both became homosexuals and both had a fear of heights, had speech problems and were hyperactive.

The odds against having twins? 90 to 1.

Only a few centuries ago, twins were considered a threat to society. Now researchers welcome the chance to study identical twins and have discovered some fascinating traits:
* There's probably a strong genetic reason for schizophrenia, since if one twin suffers from it, there's a 25 to 65 percent chance that the other one will too.
* Genes and depression are likely linked. If ̇e identical twin suffers from a certain type ̇ ̇epression, the other one does too in 74 ̇ ̇nt of the cases studied.

Fred and Frank Butler, shown during the Civil war and in old age, were at one time reputedly the oldest male bachelor twins in the world. They always lived together and never married.

Island Retreat

Pitcairn Island, a remote two-square-mile patch of land in the South Pacific, was uninhabited until mutineers landed there by chance. Now it's a treasure house of information for scientists studying such things as life-style, nutrition and inherited traits.

Nine mutineers from the *Bounty* landed there in 1790, along with six men and 12 women from Tahiti. Within a few years, 14 of the 15 men were dead—killed in fighting over the women!

The sole male survivor was English mutineer John Adams, who ruled alone on the island with the remaining women and babies galore!

The Pitcairners have lived for nearly two centuries in virtual isolation from the rest of the world. For seven generations they have married first cousins and provide a fascinating and surprising study of the effects of inbreeding:

* There are few traces of mental or physical enfeeblement.

* The men are lean and look European, the women look Polynesian.

* They have no stoves, no doctors and eat natural food low in cholesterol.

* Life expectancy is 75 years, higher than that of Europeans or North Americans.

* 12 of the original colonists had gonorrhea, which is still a major health threat and may be responsible for inherited sterility.

* The opportunity of studying this unique natural commune is fading—current population is about 80 and declining.

Out in Left Field

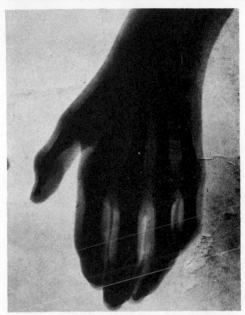

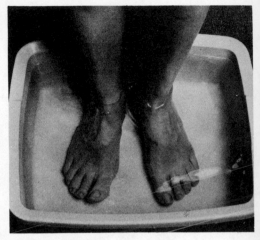

Left-handed people don't always have the best of fortune. They've been viewed with suspicion and even persecuted throughout history. Modern studies show they tend to smoke more, drink more and are more emotional and antisocial than righties.

Some southpaws win through, however. They include:

* Actors Charlie Chaplin, Terence Stamp, Betty Grable, Judy Garland, Kim Novak and Rex Harrison.

* Baseball greats Sandy Koufax and Babe Ruth.

* U.S. presidents Gerald Ford and Harry S. Truman.

* Beatle Paul McCartney.

* Tennis players Jimmy Connors and John McEnroe, and cricketer Gary Sobers.

* And Jack the Ripper!

Left Hand Down a Bit Please

A survey at Britain's Newcastle University found that right-handed people scratch with their left hand while lefties use the right.

Footloose and Fancy-free

Ask an American if his feet hurt and he'll likely say yes. An estimated 87 percent have foot problems.

Which is not too surprising, since the average North American walks 115,000 miles in a lifetime—that's the equivalent of four times round the world!

Feet are the most used and abused part of the body. Here are some of the problems:

* The fitness craze—jogging, running and dancing—has swelled the ranks of swollen feet.

* Diabetes, obesity and circulatory problems can mean trouble for your feet.

* Bunions and hammer toes are particularly common among women who wear high heels and pointed-toe shoes.

* Footwear is increasingly made of synthetic materials, which unlike leather don't "give" or let the foot "breathe." Perspiration increases the chances of infections and of smelly feet. The answer? Leather shoes or sandals.

Infertile Desert

In the Kalahari Desert of Africa, women breast-feed their babies every 13 minutes—releasing a hormone within them that stops pregnancy for two to three years!

Travel: Way to Go

You take a chance whenever you venture out on the roads or climb aboard an aircraft. In the U.S., 45,500 people were killed in road accidents in 1975—or over 120 people on every day of the year. In Britain, which had 6,520 traffic fatalities in 1976, nearly 18 people die on the roads each day.

But to compare the safety of the roads in various countries, the more significant figure is the number of deaths per 100,000 inhabitants. And here Canada, Australia and France are the places to avoid, with 26 deaths per 100,000. The rate in the U.S. is 21, while in Britain the figure is 12.

Injuries in road accidents tell a different story. The U.S. tops the list with an alarming 13.1 per 100,000, Canada is next with 9.6 and Britain is way down the list with 6.1.

More facts about road accidents:
* Your chances of surviving a car crash are excellent—98.5 percent of those involved live.
* Your chances of ending up in hospital are high: one in three accident victims is taken to hospital.
* One in three teenagers involved in a car accident dies—six times the suicide rate.
* The death rate in car crashes goes up 9.2 percent every time a famous person's death is announced on the radio!
* 6 percent of all reported motorcycle accidents are fatal—and 85 percent of those killed were not wearing a helmet. Over half had been drinking.

Belting Up

In America 70 percent of motorists on the freeway are speeding. Of these, 86 percent are not wearing seat belts.

Defective Memory

Only 60 percent of car owners send their cars in when they are recalled by dealers to replace defective parts. The rate of "return" is higher when the defect is a serious one. But when TVs are recalled by dealers, 75 percent of people respond!

Up, Up and Away

Statistics for air travel can be misleading. One major accident can distort the overall picture. One of the better ways of assessing your chances is to look at deaths per billion passenger miles. But even this isn't perfect; some airlines concentrate on short trips, with more takeoffs and landings—easily the most hazardous parts of any flight.

Between the years 1950 and 1974, Jordan had easily the worst record, with 191.30 deaths per billion passenger miles. Second was Turkey with 77.41 and third Spain with 66.93. In 1975, the Royal Jordanian Airline had 188 deaths, but none in 1976 and 1977.

The Turkish airline had 47 deaths in 1975, 155 in 1976 and none in 1977, while Spain had none between 1975 and 1977.

But with more than 30,000 people killed between 1950 and 1974 in airplane accidents worldwide, most experts say flying is safer than crossing the road.

When and How

Of the 12,004 deaths in plane crashes between 1970 and 1977:
* 20 percent occurred during takeoff.
* 31 percent during landing.
* 14 percent en route.
* 10 percent were due to collisions.
* 7 percent were caused by terrorism.

Fly regularly? Here are some airports to avoid, according to the International Federation of Airline Pilots Associations:
* U.S.: Boston's Logan airport and Los Angeles.
* Colombia, with seven airports.
* Greece: Corfu and Rhodes.
* Japan: Osaka airport.
* Iran: Tehran airport.

The One and Only You

You're unique. There's nobody quite like you in the whole wide world. And even if you wanted to disappear without a trace, chances are that you could be identified by:

∗ Your protein distribution, which can pinpoint your race, tribe and even religious community.

∗ Your microorganisms, which only you have.

∗ Your "smell prints" can identify you not only to dogs but to people as well. Clothes, body odor, diet and where you live all contribute to your smell.

∗ Your blood type, which will be one of more than two dozen.

∗ You have eight antigens in your cells, with a total of 20 million possible combinations.

∗ Your teeth can vary in a variety of ways, and dental charts have been used to convict criminals.

∗ Your signature is also unique, not only because of the way you write but also because of the pressure you put on the pen. Top security personnel in the U.S. Air Force are now identified by their writing.

Curious Cures

Willing to take anything to cure that toothache? How about crocodile scales? Here are some more weird and wonderful "cures" made from rare-animal parts:
∗ Porcupine cysts for burns: $500 per hundredth of an ounce.
∗ Whiskers of a golden cat, mixed with opium, to cure snake bites.

Many rare animals are butchered for their horns or hide to make miracle cures:
∗ Rhino horn sells in Singapore for $500 an ounce; rhino toes are $180 each.
∗ One Chinese merchant in Sumatra offered a new American car for a rhino.
∗ Japan imported 750 pounds of rhino horn in 1978.
∗ Nearly every part of the tiger has some "medicinal" use, and as a result, the tiger is nearly extinct. Only six species are left alive and the tiger population has dropped by 90 percent since 1900.

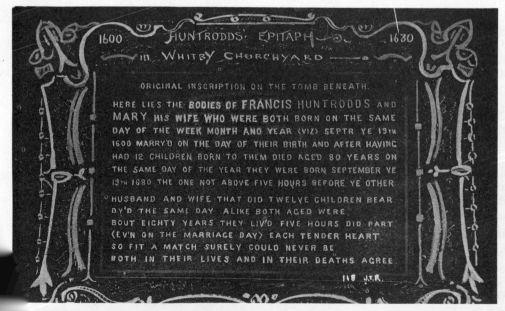

1600   HUNTRODDS EPITAPH   1680

in WHITBY CHURCHYARD

ORIGINAL INSCRIPTION ON THE TOMB BENEATH.

HERE LIES THE BODIES OF FRANCIS HUNTRODDS AND MARY HIS WIFE WHO WERE BOTH BORN ON THE SAME DAY OF THE WEEK MONTH AND YEAR (VIZ) SEPTR YE 19TH 1600 MARRY'D ON THE DAY OF THEIR BIRTH AND AFTER HAVING HAD 12 CHILDREN BORN TO THEM DIED AGED 80 YEARS ON THE SAME DAY OF THE YEAR THEY WERE BORN SEPTEMBER YE 19TH 1680 THE ONE NOT ABOVE FIVE HOURS BEFORE YE OTHER

HUSBAND AND WIFE THAT DID TWELVE CHILDREN BEAR DY'D THE SAME DAY ALIKE BOTH AGED WERE BOUT EIGHTY YEARS THEY LIV'D FIVE HOURS DID PART (EV'N ON THE MARRIAGE DAY) EACH TENDER HEART SO FIT A MATCH SURELY COULD NEVER BE BOTH IN THEIR LIVES AND IN THEIR DEATHS AGREE.

118 J.T.R.

Fingertip

What to do if your child accidentally cuts off his fingertip? The best treatment is no treatment, according to English physician Cynthia Illingworth, who found in 1974 that the finger and the nail will grow back, and be good as new in 11 or 12 weeks. But the child must be under 12, the injury must be above the first joint and no operation should be attempted.

Jumbo Risk

For every elephant born in captivity, one trainer dies—supervising the mating process....

Cancer: The Big C

Cancer. The word strikes terror into most people's hearts. It's the disease that 80 percent of Americans fear most. And it kills 400,000 a year in the U.S.; some 100,000 in Britain. Your chances of dying of cancer in the U.S. are one in five, second only to heart disease, which claimed nearly 800,000 for chances of two in five.

Some jobs seem to attract a certain kind of cancer. The highest rates:
* Stomach cancer—miners (40 percent above normal).
* Lung cancer—foundry workers (50 to 150 percent above normal) and metal workers (200 percent).
* Bladder and brain cancer—tire workers (90 percent).

The Big Killers

Cancer of the colon is the leading life-threatening cancer for Americans of both sexes, killing 50,000 people a year. Cancer of the colon causes 26 percent of the cancer deaths among women aged 50 to 54, with breast cancer next at 19 percent.

Lung Cancer

It's estimated that 122,000 Americans contracted lung cancer in 1981 and that only about 10 percent of them would live for another five years. Even being around a heavy smoker increases your chances of getting lung cancer. In a 14-year study, the National Cancer Center Research Institute found that wives whose husbands smoked were twice as likely to die from lung cancer as other nonsmokers.

Survival Odds

Here are your chances of surviving some of the types of cancer:
* Lung cancer—one in 10.
* Stomach cancer—one in 20, or less.
* Cancer of the pancreas—one in 100.
* Most others—from one in three to five in eight.

Cancer Worldwide

Your chances of dying of a particular type of cancer vary enormously, depending on where you live. The highest rates for particular types of the disease:
* Skin cancer—Australia.
* Liver cancer—Mozambique.
* Cancer of the pancreas—U.S.
* Stomach cancer—Japan, Australia.
* Cancer of the rectum—Germany, Austria.
* Lung cancer—Cuba, Scotland.

Ways to Reduce Risk

You stand a better chance of avoiding cancer if you:
* Eat less high-fat meat—beef, lamb and pork—and more fish and poultry.
* Eat more vegetables, beans and whole grains, and fruits instead of sugar-sweetened foods.
* Avoid deep-fried foods and substitute skimmed or low-fat milk for whole milk and cheese.
* Don't smoke. Two-pack-a-day smokers are at least 20 to 30 times more likely to get lung cancer than nonsmokers.
* Avoid smoke-filled places, car fumes and factory exhausts.
* Avoid needless X-rays and overexposure to the sun.
* Avoid inhaling pesticides, household solvent cleaners and other such chemicals.

Trendy Diseases

The modern world is fast, exciting—and dangerous. Here are just some of the trendy ailments that can hit you when you're in the swing of things:

* Disco felon—infection of the fingertips caused by snapping your fingers to the beat.
* Punk eye—hemorrhage brought on by New Wave jumping up and down.
* Epilepsy—brought on by the flashing lights of video games.
* Urban-cowboy rhabdomyolysis—muscle tears from riding mechanical bulls.
* Frisbee finger—caused by jagged edges on Frisbees.
* Cuber's thumb—brought on by twiddling Rubik's cubes. The cube also threatens marriages, reported a British research group in 1981. Husbands stay up all night trying to solve the cube or come to bed too tired to make love.

THE MAN WHO WAS CRIPPLED BY ONE DUEL—AND CURED BY ANOTHER!

Alexander Grailhe OF NEW ORLEANS, LA., AS THE RESULT OF AN INJURY IN A DUEL WITH SWORDS **WAS UNABLE TO WALK UPRIGHT—** IN ANOTHER DUEL WITH PISTOLS A BULLET THAT PIERCED HIS CHEST REMOVED AN ABSCESS FROM THE OLD SWORD THRUST—AND GRAILHE'S **POSTURE WAS NORMAL FOR THE REMAINDER OF HIS LIFE**

Lucky Break

Jockey Leo Cecil broke his nose when thrown by a horse during spring training at Arlington Downs, Texas, in March 1937. Three weeks later the same horse threw him again—and his nose was reset by the fall!

Charmed Life

During the 1930s "Calamity Mary" Bergere of Santa Rosa, California, was in 13 car crashes in which the cars were destroyed. She was thrown by a horse, kicked by a cow, bitten by a dog and scratched by a wild cat. She was in an Arizona cloudburst, a Florida hurricane, Kansas cyclone, Mississippi flood and California earthquake. She fell 3,000 feet in an airplane. But she escaped each and every one of the mishaps— uninjured!

It Hurts When I Laugh

Bricklayer Jim Kennedy of Galveston, Texas, fell 51 feet when a scaffold gave way in December 1937. He was knocked unconscious for seven hours, received 42 stitches in his face, head, arms and legs—but was cured of a stomach ailment that had bothered him for 15 years!

Laughter Is the Best Medicine

One way to beat fate is to laugh at it. It's true. Laughing is good exercise for the lungs and heart and maybe even for the whole body. "Healing laughter" is a medical fact; it releases endorphins, pain-killing hormones that speed healing and help arthritis, palsy and allergy victims.

* Norman Cousins, former *Saturday Review* editor, was told he had incurable cancer. He shut himself in a hotel room with comic novels and old Marx Brothers movies and laughed himself back to health. Odds against this? More than 500 to 1!
* A state prison in Missouri is now trying to rehabilitate inmates with a laugh-a-day program.
* Retirement homes and outpatient clinics in the U.S. are also using laughter for medicinal purposes.

George Cruikshank's illustration of an audience at a lecture enjoying the effects of laughing gas.

The Romantic Disease

Count Leo Tolstoi (1828–1910)

John Keats (1795–1821)

Thanks to modern drugs, tuberculosis is no longer a threat. But it ravaged the Western world in the 19th century. The chances of catching TB were high; by the 1830s it was responsible for one third of the deaths of working-class people. There were 400 deaths per 100,000 people in America, 500 per 100,000 in Europe. And:

* 13 of every 100 white prisoners in a , Philadelphia penitentiary died of TB between 1829 and 1845. The rate was even higher for blacks.
* In 1844, all 78 boys and 91 out of 94 girls in an English workhouse had TB.
* In the Hôpital de la Charité in Paris in the early 1800s, 250 out of 6967 people died of TB.

And it actually became fashionable to be consumptive! TB and genius became linked: "Everyone was consumptive and especially the poets," wrote Alexandre Dumas père (1802–1870). "It was good form to spit blood from sheer emotion."

It was the tragic early death of the English poet John Keats (1795–1821) that sparked the rage. His poems were considered the product of the hectic, fevered imagination of the consumptive.

Paleness became all the rage. Lord Byron said he would like to die of consumption because the ladies would say how interesting he looked. People took lemon juice and vinegar to kill their appetites and stayed out of the sun to look pale. Many probably died for that very reason—of consumption.

Others who died of consumption:
* American Henry David Thoreau (1817–1862).
* Scottish novelist Robert Louis Stevenson (1850–1894).
* German poet Friedrich Schiller (1759–1805).
* German composer Karl Maria von Weber (1786–1826).
* Italian violinist Niccolo Paganini (1782–1840).

In Sickness and in Health

Henry David Thoreau (1817–1862)

Niccolo Paganini (1782–1840)

Johann Friedrich von Schiller (1759–1805)

It's one of the ironies of fate that creativity may be boosted by sickness and suffering. Many of the world's great artists did their best work when they were ill. For example:
 *Russian novelist Leo Tolstoi had syphilis and gonorrhea in his youth. He constantly contemplated suicide.
* Impressionist painter Vincent van Gogh was probably a manic-depressive and an epileptic. He committed suicide at the age of 37.
* Nineteenth-century poet Lord Byron had a

club foot, epilepsy and had to live on a starvation diet due to a hereditary glandular imbalance.
* Pianist and composer Frederic Chopin had TB, and his 24 Preludes were written during a bad spell of pulmonary illness.
* Catholic mystic St. John of the Cross wrote his famous poems while he was in jail and being tortured by the Spanish Inquisition.
* English poet A. E. Housman never wrote anything unless he was sick in bed!

This Won't Hurt Now

The placebo, or harmless sugar-pill, is more powerful than you think. Provided you don't know what's really in it, chances are it will do you good. Placebos depend on faith. Here are some examples of faith in action:

* A team at the University of California studied people who had just had their wisdom teeth extracted. Some were given morphine, others just received a saline solution. Yet a third of the patients in both groups reported that the pain was going away.

* Placebos have cured heart and blood pressure problems, digestion complaints, colds, ulcers, seasickness, vertigo, hay fever — in fact almost every ailment studied.

* Valium may be nothing more than a placebo, especially if taken in its customary small doses.

* In a study of angina patients, some had major blood vessels cut and tied off, others merely had their chests nicked and were sewn up again. Those who had the sham surgery did better than those who had had the real operation.

MEDICINE MEN OF THE MAMBUNDU TRIBE, AFRICA, BELIEVE THEY CAN DISPEL ANY ILLNESS **SOLELY BY DANCING**

Ancient Remedies

Some believe the placebo effect was responsible for the success of doctors in the old days who treated people with some of the most bizarre medications imaginable — including crocodile dung, eunuch fat and moss scraped from the skull of a hanged criminal….

THE **MOST RADICAL TREATMENT FOR THE COMMON COLD** KINGS of the Shilluk Tribe in Africa **FOR CENTURIES WERE EXECUTED IF THEY CAUGHT A SLIGHT COLD** THE TRIBESMEN BELIEVED THEIR FORTUNES DEPENDED ON THE MONARCH'S STATE OF HEALTH – SO THEY STRANGLED ANY RULER WHO WAS NOT IN PERFECT HEALTH

Cold Comfort: The Kissing Doesn't Have to Stop

What are your chances of getting a cold from a kiss? Negligible, say the experts. Colds are spread from the nose, not the mouth.

So even a three-minute kiss won't necessarily leave you with a cold — though it may take your breath away!

A-a-a-choo!

Your chances of getting a cold during the year are one in four — higher if you have children. The average person gets 140 colds in a lifetime — which is nothing to sneeze at.

Man's Best Friend

Feeling down? Get a pet—it'll probably do you a world of good. In a study of inmates at a hospital for the criminally insane in Lima, Peru, many prisoners tried to commit suicide. But none attempted to kill themselves in wards where pets were kept.

Doctors say pets keep people busy, thus relieving tension and anxiety. They make you feel less isolated and better all round.

Americans agree: they keep 70 million pets, one for every three people. And in one study of patients with heart disease, there were 11 deaths among the half who had no pets, only three among those who did.

PETER BARR
of St. Faiths, England

TOTALLY BLIND FOR 10 YEARS SUDDENLY REGAINED HIS SIGHT *WHEN HE POUNDED HIS HANDS TOGETHER IN AN ARGUMENT*

ERASMUS
(1466-1536) MOST FAMOUS HUMANIST OF THE RENAISSANCE, WAS ALWAYS STRICKEN WITH A FEVER *AT THE SIGHT OF LENTILS OR FISH*

Doctors: Diagnosis Is Bad

Happy with your doctor? Probably not—60 percent of all Americans are losing faith in their doctors, according to a recent American Osteopathic Association poll.

Other findings:
* 72 percent said fees were too high.
* 50 percent thought the health system discriminated against the poor, and half thought the overall system was merely fair to poor.
* 50 percent said doctors didn't want to know their patients and didn't take the time to explain things properly.
* 35 percent thought doctors overprescribed drugs.
* 72 percent said there was a shortage of doctors.

Actually, the U.S. has 16.5 doctors per 10,000 inhabitants, while Britain has 15.3 and Canada 16.6, according to World Health Organization figures.

Death Takes a Holiday

A frustrated Louis Pasteur took a vacation in the midst of experimenting with deadly cholera germs. When he returned, he discovered his cultures had become sterile and he found infected fowls were thereafter immune to new germs. The great French doctor drew the correct conclusion—weak doses of germs promote the body's ability to produce antibodies—and his

chance discovery became the foundation of all modern techniques of immunization. We can thank Pasteur for saving the world from smallpox, measles, diphtheria, whooping cough and scarlet fever.

False Teeth: An Ancient Secret

An ancient people living on the Italian peninsula, before the Romans, developed by trial and error dental procedures that are still in use.

The Etruscans, concerned about teeth lost in battle or through disease, perfected techniques of duplicating them from bone and ivory. Sometimes they even used teeth pulled from cattle.

The teeth were set in partial plates and fastened to the remaining teeth in the mouth with gold bands. These ancient ideas have come down to us more or less unchanged despite the fact they were developed in 700 B.C.

Oddly enough, Europeans and North Americans in later centuries abandoned the marvelous Etruscan principles for a while and used backward methods. George Washington, for example, was proud of two crude wooden molars jammed in between his good remaining teeth.

No Laughing Matter

A wilted flower gave us one of the world's best-known anesthetics.

Although ether was first discovered in 1549, nothing was done with it until 1923 when a British doctor named Lockhardt noticed his carnations wilted when they were exposed to it.

He wondered what the gas would do to a human body and tried it on himself. He briefly passed out; when he awoke, he had proven a practical use for the first modern anesthetic.

Strangely, Valerius Cordus observed the same phenomenon in 1549, but made no more of it. And in 1842, Dr. Crawford Long revived a patient who had sniffed too much of it, but no one understood its potential as an anesthetic and the gas was used only at parties as a titillation for young socialites.

Cleaning up the Act

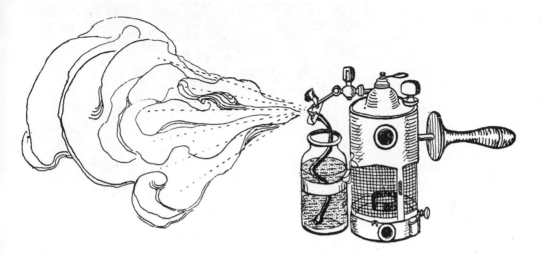

Joseph Lister's carbolic spray.

The great surgeon Lister, sick of bloodied scalpels and filthy operating rooms, decided to wash everything before he operated. That chance decision brought us the first crude sterilization procedures.

The British Medical Association of the early 1800s scoffed at this "mad doctor" who maintained infection was caused by dirty instruments. Everyone knew it was caused by noxious vapors in the air.

But Lister persisted and gradually the medical establishment came around. He helped wipe out puerperal fever, which can affect mothers who give birth in unclean conditions. Before Lister's discovery, most women had delivered their babies at home rather than go to filthy hospital delivery rooms.

First You Walk the Dog ...

Everyone knows Banting and Best discovered insulin in 1921, but one of the world's greatest medical breakthroughs really depended on some chance tests on dog urine.

Two researchers named Mehring and Minkowski noted that flies gathered around dog urine, and they wrote a paper in 1871 speculating that it contained sugar. Banting heard about their tests and eventually drew the conclusion that diabetes was connected with a chemical which controlled the sugar levels in the blood.

From there, it was a short but tricky step to isolating the hormone, insulin, in the pancreas, and then synthesizing it.

THE MAN WHO WAS DEAD, FOR 3 DAYS!
SAI BABA
(1856-1918) of Shirdi, India,
WAS PRONOUNCED DEAD IN 1886,
WITH BOTH CIRCULATION AND
BREATHING STOPPED COMPLETELY.

AS PREPARATIONS FOR HIS
FUNERAL WERE BEING MADE 3
DAYS LATER, IT WAS OBSERVED
THAT HE WAS BREATHING--AND
HE LIVED ANOTHER 32 YEARS

The Eyes Have It

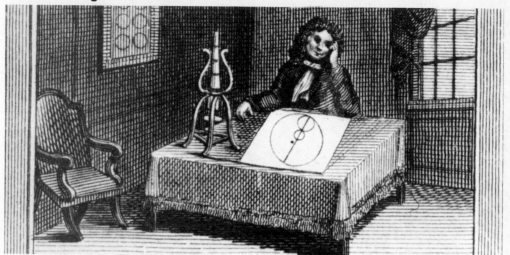

Frustrated in his attempts to see bacteria, a Dutch doctor invented the first microscope in 1674.

Anton van Leuwenhoek probably did not realize the importance of his discovery until he decided by chance to send specimens to the London Royal Society. Doctors were amazed at the quality of his germ sections and asked Van Leuwenhoek how he had done it.

He explained that he had made tiny lenses from glass by blowing bubbles and letting droplets form. The drops were then ground and held with tweezers over the culture.

Danish Blue

A Danish physician was a little clumsy—and gave us a medical breakthrough.

Dr. Gram was wondering how to isolate and identify some germ cultures when he accidentally spilled iodine on them. Not only did the cultures remain unharmed, but they were also clearly identifiable.

These procedures are still used today in modern forensic labs.

An Inside View

A war injury to a French soldier gave us the chance discovery of the digestive process.

Dr. William Beaumont treated a soldier who had had part of his stomach shot away. The wound was slow in closing and Beaumont was able to observe the digestive process as it actually took place in a living body.

His early observations (and an unknown soldier) gave us the groundwork for the study of the stomach.

Double Chance

Two of the world's greatest medical breakthroughs came from a barber's apprentice who by chance deviated from the accepted practice of the day.

Ambrose Pare, like everyone else in the early days of medicine, used boiling oil to cauterize open wounds. But one day he ran out of oil and decided to use a clean dressing instead, giving us the first bandage.

Later, he developed the modern art of tying off major blood vessels after amputation. The custom at that time was to burn the ends of the vessels, but Pare felt sorry for his patients because they had already suffered enough pain.

Medical Miracles

Chances are our lives will be a lot better in the future because of some fascinating new medical miracles. But the past is also full of medical stories studded with chance, fate, coincidence and good old-fashioned luck.

Come with us now as we examine miracles of the future and miracle medical cases of the past.

A Vital Center

Is it chance, location or the long Canadian winter? No one knows for sure, but Toronto is considered one of the major centers of medical discovery.

* Dr. Frederick Banting and Dr. Charles Best discovered insulin in Toronto in 1921 and won the Nobel Prize for medicine.
* The first electron microscope in North America was built at the University of Toronto by Eli Burton.
* The first solid-state computer was developed at Ferranti Electric Ltd. just after World War II by Dr. Arthur Porter.
* The city's Institute for Aviation Medicine designed the first human centrifuge and pressure suits for pilots during World War II.
* The cobalt bomb for treating cancer victims was developed by Dr. Harold Johns.
* Pablum was created by Dr. Allan Brown at the city's Hospital for Sick Children.
* Life-saving surgery techniques for blue babies were developed by Dr. William Mustard and are now used worldwide.

A Star Is Born

A chance examination of a starfish has given us new breakthroughs in research into bone growth.

Dr. Arthur Fontaine, a Canadian biologist in British Columbia, noted the starfish skeleton is made up of a porous calcium carbonate. He used it as a base to grow successfully three kinds of bone cells.

He took his research further by dropping the material in holes in the leg bones of rats. The bones regenerated even faster than under normal circumstances!

Swatting Them Out

The cursed blackfly may soon be combated thanks to another summer pest.

Scientists have found that a bacterium on dead mosquitoes inhibits the growth of blackflies. They are now working on ways to synthesize it and paralyze the blackfly's reproductive system.

The World Health Organization says the new strain is almost ready to be produced by pesticide companies.

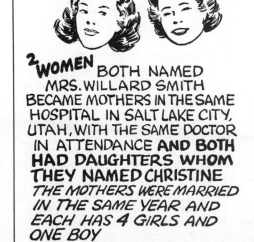

2 WOMEN BOTH NAMED MRS. WILLARD SMITH BECAME MOTHERS IN THE SAME HOSPITAL IN SALT LAKE CITY, UTAH, WITH THE SAME DOCTOR IN ATTENDANCE AND BOTH HAD DAUGHTERS WHOM THEY NAMED CHRISTINE THE MOTHERS WERE MARRIED IN THE SAME YEAR AND EACH HAS 4 GIRLS AND ONE BOY

Call Me in the Morning

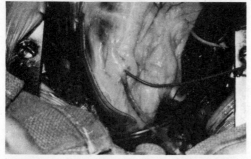

Chances are your future heart attack could be treated over the phone.

A new machine plugs into an ordinary telephone and lets a doctor diagnose heart conditions even though his patient may be thousands of miles away. The device is carried in a simple attache case and if a cardiac-prone person feels an attack coming, he can attach a sensor to his chest and plug the machine into a phone.

The doctor monitors the patient's condition and, if necessary, prescribes the correct dosage of drugs.

Not Tonight Dear, I Have a Headache

A New York firm is trying to take all the chance out of becoming pregnant.

Intersonics Inc. has developed a watch-sized device called the Fertilitron which monitors minute changes that occur in a woman's body when she ovulates.

By pressing the watch against her wrist, a woman can obtain a reading from a micro-sensor. If the light's green, she's fertile. Red means no ....

Bionic Blood

The odds are good that synthetic blood will soon be here.

Doctors in Japan have already developed a type of blood made from liquid teflon. It can be given to anyone and keeps for at least two years. It still cannot clot like real blood, but doctors are working on it....

It Beats a Bloody Mary

The primrose plant, of all things, may improve your chances of surviving a hangover.

Oil from the plant contains a chemical called linoleic acid, which is lost from the body when a person drinks a lot of alcohol. One of the best ways to ingest it is to boil up a primrose plant and drink it like tea.

Stick 'em Up

A new miracle glue may soon replace painful stitches.

A Canadian doctor named Khadry A. Galil of the University of Western Ontario is testing a new medical glue that dries to form a thin plastic skin which closes wounds in a second and stops bleeding.

Two CCs of Chilies, Nurse

Chances are the chili does more good than harm.

For years, medical researchers have thought the hot pepper produced indigestion, stomach ulcers and cancer, but now they've changed their minds.

New studies show eating chilies induces the body to produce endorphins, a natural painkiller with the same effect as morphine. And chilies also make the face perspire, which helps the body to cool itself in torrid climates.

Tanks for the Memory

Chances are that a few years ago you never would have considered lying in a pool of warm water to cure tension.

But medicine of the future includes the isolation tank where one hour in pitch darkness lying in blood-warm water is said to equal eight hours of normal rest. The tanks induce an alpha state of deep relaxation and are being used to cure jet lag and nerves.

RUDOLPH BUGUARDINI (1864-1912) of FLORENCE, ITALY, ENJOYED PERFECT HEALTH THROUGHOUT HIS LIFE -- *EXCEPT THAT HE RAN A FEVER ON EACH OF HIS 48 BIRTHDAYS*

Bust or Boom?

A New Jersey hypnotist says the chances are good you can think your way to bigger breasts.

He gives women a posthypnotic suggestion that their body weight will shift upward and enlarge their chest size. And he offers a handy cassette tape to take home to continue the therapy.

New Help for the Habit

A heroin addict once had almost no chance.

If he was addicted, doctors used to prescribe methadone, an artificial drug which also had addictive properties.

But now a new drug, naltrexone, has been developed. It neutralizes the effects of even a large dose of heroin, giving addicts time to readjust and shake the habit.

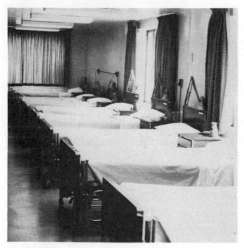

And Now, the Research

Modern medical science moves so quickly it is almost impossible to keep up with its discoveries. But chances are good you will soon benefit from these recent developments:
* A vaccine to guard against meningitis. Discovered in Finland, it has worked in tests to date.
* A new artificial kidney has been effectively demonstrated to save kidney victims from dialysis treatments. It weighs about eight pounds and is worn around the patient's waist.
* A laser beam treatment that seals off blood vessels in the inner eye and prevents them from bursting and causing blindness.
* A new chemical called cemetidine which inhibits acid secretions in the stomach and enables ulcer victims to eat normally.
* A chemical that dissolves gallstones.

The Deadly Toast

British novelist Arnold Bennett was sure of the drinking water in Paris. In 1931 he drank a toast to prove it was safe. But the drink proved deadly and Bennett died, in the city he loved, of typhoid contracted from the germ-filled water!

Southern Fried Gel

Chicken feet may soon provide a new burn treatment.

Scientist Harry Robertson discovered the process while working in his lab in Maryland. He accidentally burned himself over a Bunsen burner and by chance reached for a new gel made from chicken feet.

His untested process worked. The pain stopped immediately and Harry's burn healed without a scar. The gel, called Revital, is now being market-tested. Harry claims it regenerates nerve and muscle tissue, heals third-degree burns, and eliminates acne scarring, skin ulcerations and bedsores.

Sex Changes

Girls who turn into boys have been discovered in the Philippines.

The children are the victims of a rare chance gene which inhibits the development of the male hormone. They look like girls until puberty and then turn into boys and later men.

Thirty-seven children have been born this way already. They have vaginas at birth, but around 13 they are flooded with male hormones and grow normal-sized sex organs.

Oddly enough, the society they live in treats them like girls until the change occurs and then naturally accepts them as males.

LITTLE MISS **XAO CHING-PING** OF SHANGHAI

WAS SUDDENLY CHANGED INTO A BOY DURING A TERRIFIC THUNDERSTORM

(THE INDESTRUCTIBLE MAN!
SGT. NICHOLAS RIGOPOULOS

SHOT 9 TIMES THRU THE HEAD - STILL LIVES!

HE WAS CAPTURED BY THE COMMUNISTS IN PARNASSIS, GREECE, AND SHOT BY A FIRING SQUAD - LATER HE WAS FOUND TO BE STILL ALIVE AND WAS EXECUTED A SECOND TIME! — AND AGAIN A THIRD TIME — BUT WITH 9 BULLETS THRU HIS HEAD HE WALKED 2 MILES TO A VILLAGE AND WAS RESCUED

Unlucky Breaks

William Russell was a private in the American army serving in Sicily when he fractured his leg in heavy fighting on September 22, 1943. One year later he was in Lawson Hospital in his home state of Georgia when he turned over in bed and broke the same leg again on the same date, September 22. Exactly two years later, on *September 22*, 1946, he fell and broke the leg for a third time in the same place!

The Christmas Gift of Health

A 71-year-old gold prospector in Mirandanao, South Africa, fell 50 feet from a rocky cliff in 1935, suffering a dislocated vertebra. The pain was so intense that Frederick David Burdett had to walk with a cane. He then stumbled on Christmas Eve, falling in front of a heavy oxcart. The cart ran over the damaged vertebra and Burdett's dislocation was cured. He never suffered a backache again and lived until 1942.

The Duel of Death

Colonel Acland of Selworthy, England, fought and won eight duels in the 1700s, but was warned by a soothsayer that his ninth fight would result in death. The brave colonel scoffed at the advice and fought and won another duel on November 8, 1778, escaping without a scratch. The chilly, wet English morning took its toll, however, and Acland caught a cold and died—exactly four days after that fateful ninth duel.

The Gift of Love

In 1915 Jean Castel, a 15-year-old French youth, fell and suffered total amnesia while fleeing from German soldiers in the Great War. Nine years later, still unable to remember his first 15 years of life, Castel met a beautiful English girl and fell in love with her. When she finally told him her name—also Jean Castel—his amnesia vanished and he totally recovered his memory!

Deadly Prescription

Francois Vautier stunned the French medical world in 1631 when he became the first to use antimony powder in all his cures. So convinced was Vautier of the values of the bluish-white element that he included it in every prescription he wrote. But alas, Vautier died of antimony poisoning in 1652. The element is now used in explosives....

The Sting of Life

August Pfeiffer was only 5 when he was pronounced dead in 1645 after a fall from a tree. The boy was about to be buried when a bee stung him, causing him to flutter his eyelids. He remained in a coma for 70 days; when he recovered, he vowed to learn 70 languages. August died at the age of 58, the most famous linguist in Germany, only days after he had learned his 70th tongue.

50 Years of Accidents

John Foard of London, England, broke more than the laws of chance. He broke one of his legs every year for 50 years until he died in 1910, with the dubious accomplishment of being the most accident-prone man in history.

THE REV. JOHANNES OSIANDER
(1657-1724) of Tübingen, Germany,
WAS KNOCKED DOWN
BY A WILD BOAR,
HAD HIS HORSE FALL ON
HIM DURING A FLOOD,
WAS SHOT AT BY BANDITS AND
BURIED UNDER AN AVALANCHE,
WAS BLOWN INTO THE RHINE
RIVER BY A BLIZZARD,
CRUSHED BY A FALLEN TREE,
WAS SHIPWRECKED – AND EVEN
WAS RUN OVER BY A SHIP
HE ESCAPED UNHARMED
EVERY TIME

The Last Song

Famed songwriter Vincent Youmans had just reached the top of his profession with "Tea for Two" when he dreamed that his mother told him he was dying of tuberculosis. He was in full health, yet half an hour after waking, Youmans had his first hemorrhage and eventually died of TB.

Ice Follies

Dean Haase and his wife loved to spend New Year's Day skating on a lake in their hometown of North Platte, Nebraska. But Haase fell on the holiday in 1954 and broke his left leg. It healed in time for him to be back on the ice on New Year's Day, 1955, at which time his wife fell and, chance or not, broke her right leg!

A Call to the Bar

A 24-year-old carpenter in Los Angeles defied more than the laws of chance.

In July 1981 Michael Melnick fell 10 feet through the second floor of a construction site and impaled his head on a steel reinforcing bar. The metal rod pierced his skull and penetrated the lower part of his head, coming out between his eyes.

Astounded surgeons at Westlake Community Hospital, where Melnick was brought in, finally decided to pull the bar out by hand rather than operate.

Incredibly, the bar hadn't damaged his eyes, spinal column or brain. It didn't even sever any arteries. The only damage was to his nose and tear ducts.

Doctors had no idea how he lived because there is no room in the head for the bar to have passed through without killing him. But Melnick is alive today to prove them wrong.

He Lived to Fight Again

Col. James Gardiner was rallying his troops during fierce fighting at the Battle of Ramillies in the Spanish Netherlands on May 23, 1706, when a rifle ball passed through his open mouth while he was shouting commands.

By chance, the ball missed his teeth, tongue, gums and palate and passed out through the back of his neck. The hole in his neck remained for 39 years until the colonel died of other wounds in the Battle of Prestonpans.

THE STRANGEST SERIES OF DUELS IN HISTORY!

HENRI TRAGNE of Marseilles, France, FOUGHT 5 DUELS

IN THE FIRST 4 EACH OF HIS OPPONENTS FELL DEAD BEFORE A SINGLE SHOT WAS FIRED – *AND IN THE 5th TRAGNE HIMSELF WAS FATALLY STRICKEN IN THE SAME MANNER* (1861 - 1878)

Say Aaah

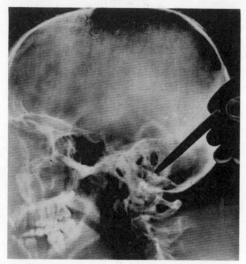

X-ray showing a bullet which had lodged in a man's skull, and which was successfully removed by surgeons.

And now, a few medical miracles involving the mouth—and chance:

∗ John Hudson of New York was shot in the mouth by a bandit in 1933 and escaped without a scratch. The bullet bounced off his dentures.

∗ Patrick Mackilwyan of Britain grew three new teeth at the age of 107, and when he was 110 his snow-white hair started to turn blond. He died at age 111 in 1663.

∗ Madame Isabella Oreille, a noted pianist in Paris at the turn of the century, was born without ears. But she could hear through her mouth. By chance, she married a man named Oreille—the French word for ear.

∗ Gene Turner of Hope, Idaho, was shot through the mouth and the bullet clipped off a tonsil and emerged from the back of his neck. Two weeks after the accident he was out of hospital.

∗ Children born in Italy in 1348, the year of Europe's Great Plague, grew only 22 to 24 permanent teeth instead of the normal 32.

And More of the Same

A steelworker bent over his workbench in 1933 to pick up a fallen tool and chance stepped in.

Paul Kovski was hit by a dirty, scaly bar coming out of a tubing machine and before co-workers could save him, 24 feet of the steel rod was driven through his head, entering at the neck and coming out the forehead just over the right eye.

But luck was on his side. Kovski never lost consciousness and after the bar was removed—again by hand—made a complete recovery and returned to work at the steel plant in Hammond, Indiana.

See Here

A chance knock on the head restored a man's sight after 14 years.

William Passmore of Dartington, England, was shoveling coal when he hit his head. It was only a minor bruise but when Passmore awoke the next morning, he could see again.

Doctors had originally diagnosed his blindness as having been caused by cataracts.

A woman displays a pair of surgical forceps which was accidentally left in her stomach 20 years before.

Hear here

A 9-year-old boy who had been deaf for three years was cured by a one-in-a-million chance.

Floyd Mills of Los Angeles had become deaf and dumb in 1950 after contracting diphtheria. During a ride in the family car one night, he bumped his head against the dashboard.

The family rushed him to hospital, but on the way, Floyd shouted: "I hear, I hear noises!"

Animal Tale

A European prince had a lion as a medical indicator.

Prince Eugene of Savoy kept a lion in his private zoo in Vienna and, by chance, every time the prince became sick, the lion also got ill.

The lion roared at midnight on April 21, 1736, and at the same moment its owner died in his city palace. The next morning the lion was found dead in its cage.

BROTHERS REUNITED AFTER **30** YEARS BY THE **SMELL** OF AN **APPLE**

A TRAVELER STANDING IN THE R.R. DEPOT IN STILLWATER, OKLA. WAS EATING AN APPLE, WHEN A STRANGER APPROACHED AND SAID, "THAT SMELLS LIKE A NORTH CAROLINA APPLE." "IT IS", SAID THE TRAVELER, "I'M FROM NORTH CAROLINA." "SO AM I", SAID THE STRANGER — AND THEY TURNED OUT TO BE BROTHERS WHO HAD NOT MET IN 30 YEARS!

What Took It So Long?

Guy Gilbert of Santa Ana, California, swallowed a needle in 1888. Seventy-six years later, it was removed from his leg.

A Child's Luck

A 19-month-old baby in Illinois had more than chance on her side in 1956.

Karen Wahl had crawled out of her backyard and onto a building site next door when a 14-ton truck ran over her. The baby escaped without a single broken bone despite the fact she was covered with treadmarks.

That One Looks Familiar

What are the odds of this ever happening?

During World War II, Americans donated 13.3 million pints of blood in Washington, D.C., and one donor had his life saved by his own blood.

Harry Starner, who had been wounded near Tarawa, looked up from his hospital cot and saw his own name as the donor on the label of the plasma bottle.

Ironies of Chance

The pioneer in blood plasma research died a death of cruel irony.

Dr. Charles Drew of Washington bled to death in 1950 of injuries received in an automobile accident.

He saved thousands of lives during the war because he developed ways of preserving plasma and storing large quantities of blood.

Bang, Bang, You're Alive

A calvary officer had his life saved by a bullet.

Frank Harrington was one of 50 men attacked by 500 Cheyenne Indians in 1868 when he was hit in the forehead by an arrow.

He could not extract it, but a glancing rifle shot hit it and knocked it free.

Brainy Breath

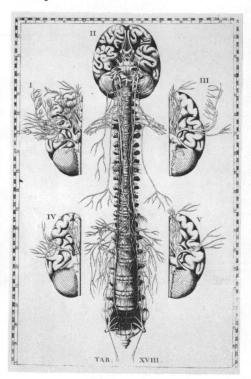

Chances are you never really stopped to think about your brain, even though the brain is doing the thinking. Or how about your nose? Consider now these important parts of the body and some chance research that affects us all today:

* Paul Broca of France pioneered phrenology in the 1880s by convincing scientists that intelligence could be measured by the size of the brain and skull. The bigger the head, the smarter the person. But alas, he was proven wrong. Einstein's brain, for example, was slightly smaller than normal and Neanderthal man had a huge brain.

* Chances are men have more brains than women. The brain mass of the average man weighs three pounds while a woman's weighs two.
* Were we once snakes in the grass? Chances are we were, according to scientist Carl Sagan, who traces the development of primitive parts of the brain back to reptiles and dinosaurs. He says man's forebrain is the same as that of dolphins and whales, which also communicate in a spoken language.
* Chances are good you can control your brain's operation through breathing. Yoga masters have been practicing it for centuries, but modern science now backs them up.

We alternate nostrils during regular breathing, and some yogis feel we should do this consciously, sending air to the appropriate hemisphere of the brain for active and passive conduct.
* If you have a nose blockage due to a deviated septum, chances are you are prone to stress-related diseases.
* Scientists have proven that menstrual cramps are related to inflamed nasal passages.
* If you're a newlywed, chances are good you'll inflame your nose during lovemaking.

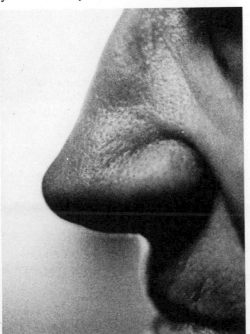

Fast Relief?

Chance can be cruel, despite the marvels of medical science and the generosity of rich countries. Food to the starving of the world is the goal of countless campaigns, but we find:
* A shipment of supplies to the stricken in Biafra included a consignment of evening gowns.
* Refugees from floods in East Pakistan received a supply of brassieres.
* Indian Muslims in the same flood received tins of pork, even though they are forbidden by religion to eat it.
* Refugees in the African area of Welo were rounded up and sent to camps to receive food, but hundreds died because there was no proper hygiene in the tents set up to save them.
* 4,800 cases of Metrecal, a weight-loss drink, were sent to Vietnam in a famine relief program.
* During the Biafran crisis, thousands of tons of food were left to rot in ports because the government had no means of distributing it.
* Powdered milk and baby formula given to the starving have been proven to make chances of survival worse. They discourage breast-feeding and spread germs from unsanitary water and bottles.

Splash

If you have to fall into water, make sure it's cold.

Latest medical tests show that people who fall into cold water survive immersion up to 38 minutes because the body cuts oxygen flow to outside tissues and sends it instead to the heart and lungs.

Eau de Vie

Sir Hugh Acland of Killerton, England, was pronounced dead in 1770 but revived when a footman sitting with the "body" poured a glass of brandy down his throat! He lived another 18 years!

The Whole Tooth

Doug Pritchard of Lenoir, North Carolina, consulted a doctor when his left foot became painful—and found he had a tooth growing there!

Hale and Hearty

Born with a hole in the wall of his heart, later corrected by surgery, Gary Douglas Gibbs of Bossier City, Louisiana, has a birthmark directly behind his heart on his back—in the shape of a heart!

Free Fall

Sgt. Nicco Alkemade, an RAF rear gunner, bailed out of a plane in Westphalia, Germany, on March 24, 1944. He had no parachute and plunged 18,000 feet, landing unhurt in a snowbank only four feet deep!

The High Life

It seems like a glamorous profession, with lots of travel and the chance to meet new people every day. But being a flight attendant on a major airline is no bed of roses. They suffer from:
* Insomnia from the time changes and strange hours they work;
* Constipation and stomach troubles from irregular meals;
* Loneliness;
* Tiredness so acute they sometimes stumble into walls.

Sick to Death of Work

Chances are good that hard times bring on disease and death.

A study at Johns Hopkins shows that a rise in unemployment brings a corresponding rise in sickness. When the unemployment rate went up 1.5 percent in the U.S., there was an increase of 20,240 deaths from heart and kidney disease, 495 deaths from liver cirrhosis, 920 suicides and 684 homicides.

In the 1.5 million unemployed, there were 36,887 deaths, 3,340 admissions to state prisons and 2,227 admissions to mental institutions.

In God We Trust

What are the chances of survival for Jehovah's Witnesses who refuse blood transfusions?

Better than average, to judge by records in Houston, Texas, hospitals over a 20-year period. Of 542 operations performed without benefit of transfusion, all were successful.

Physicians there are now wondering if most operations use too much blood, which involves the risk of transmitting disease.

Large Oversight

Lt. Thomas Falla, of the British 12th Infantry Regiment, was injured during the siege of Seringpatam, India, on April 6, 1799, and was treated for a surface wound. But after his death, a 26-pound cannon ball was found inside his thigh!

DURING A FIGHT —

A KNIFE 5 INCHES LONG WAS DRIVEN INTO THE FOREHEAD OF A GILBERT ISLAND NATIVE

With the Knife in his head this native killed his man!

THE KNIFE REMAINED IN HIS FOREHEAD 3 DAYS — BUT HE SUFFERED NO ILL EFFECTS 1934

Good Insurance

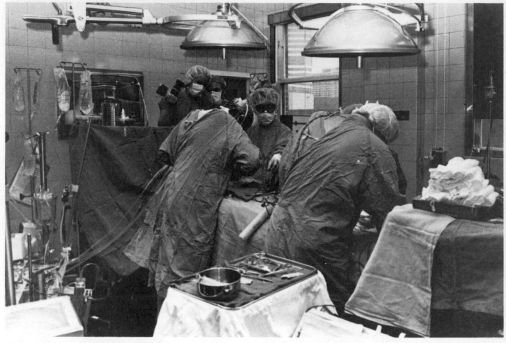

If you're a North American covered by some kind of medical program, chances are you'll undergo twice as many operations as someone without insurance.

Germ Warfare

War can be dangerous to your health. But disease can be more so—even on the battlefield. The Union army lost more men as a result of disease than it did in combat during the Civil War. The same happened to the U.S. forces in the Spanish-American War and World War I. And in the Mexican War, the U.S. lost 1,733 men in battle—and 11,550 to "other causes."

Man's Best Friend?

Patting your dog may increase your chances of getting multiple sclerosis.

A British journal recorded that of 29 average patients, 3 had dogs and 2 had cats. Another doctor tested 50 MS patients and found that an incredible 46 had had close contact with a family pet just prior to contracting the disease.

DEATH COMES FOR FARMER CODY IN A WEIRD AND FREAKISH MANNER

THE STEERING WHEEL OF A MOTOR TRUCK BROKE, PUTTING CAR OUT OF CONTROL AND CAUSING IT TO CRASH INTO A 4,000-VOLT HIGH POLE LINE
A LIVE WIRE FELL ON A METAL ROAD SIGN WHICH CHARGED A BARBED WIRE FENCE WHICH SET A PATCH OF GRASS ON FIRE IN FRONT OF THE CODY HOME
FARMER CODY THREW A PAIL OF WATER ON THE BLAZE—THE WATER ACCIDENTALLY HIT THE FENCE
AND CODY WAS ELECTROCUTED!

Hit Us Again

Chances are you'll be healthier if you learn to meet stress head-on. And new studies show the ability to handle stress can make better people.

A prison rehabilitation program in California uses criminals who have committed violent crimes or have a record of violent behavior. Just before they are released, they are provoked, cursed and humiliated, given degrading jobs and told they are worthless.

Authorities feel if they can handle this, they are ready for anything. And statistics show that only 4.5 percent ever end up in prison again—about 10 times less than the usual rate.

Frankly Amazing

* Two women named Irene Smith, both 47-year-old blondes, entered Saint Joseph's Hospital in Tampa, Florida, on June 10, 1951. Both had the same type of operation on the same morning by the same doctor. Their husbands were both called Frank!
* Biago di Crescenzio of Rome was rushed to hospital by a passing motorist after his car skidded off the road and hit a tree on May 20, 1974. He was being transferred to another hospital when the ambulance hit an oncoming car! Again, he was rushed to hospital by a passing driver. Placed in another ambulance, he was killed when the ambulance skidded into an approaching car!

Worst Cure

Chances are this takes the prize for the worst medical treatment of all time.

According to history, it was a stroke that killed Charles II, king of England, in February 1685. But maybe that wasn't what did him in:

* On the morning of his stroke, 12 physicians were summoned, and they immediately embarked on a treatment designed to purge the poisons from his body.
* They relieved him of a quart of blood, fed him substances to make him vomit and purged his intestines with a 14-ingredient enema.
* They then shaved his scalp and singed it with red-hot irons, filled his nose with sneezing powder and covered him with hot plasters.
* When that did not work, they gave him 16 more enemas and dabbed his face with resin and pigeon droppings.
* They even tried pearls dissolved in ammonia and powder from a human skull. On the fifth day, Charles died.

Fascinating Facts

Chances are you didn't know:
* More than 100 million people in America regularly read the comic strips and the peak age for Sunday comics is 30 to 39.
* A head-on collision between two cars occurred in Redruth, Cornwall, in 1906, which might not seem like much except they were the only cars in town.
* The most common last names in Britain: Smith, Jones, Williams, Brown, Taylor, Davies, Evans, Thomas, Roberts and Johnson.
* Wallace Williams of Charlotte, North Carolina, accidentally dropped his watch from a plane at 2,000 feet and found it in his own backyard—still running.
* A tortoise marked by Robert Brown in 1887 with his initials was found 64 years later in Centertown, Kentucky, by his son-in-law, O. J. Porter.
* Jim Hunkin fell into the water in Mevagissey, England. Another Jim Hunkin threw him a life preserver and a third Jim Hunkin pulled him into a boat and saved him. None of them were related!
* In 1979 in the United States, there were 3.4 million live births, a 4 percent increase over the previous year. There were 1.9 million deaths during the same year, a rate 1.1 percent lower than in 1978.

AMAZING SURGICAL OPERATION
"CHIEF" COUZZINGO, Sioux Indian, Age 70, of Oxford, Ohio
TOOK AN ICE PICK AND A SCREW-DRIVER AND FASTENED A BROKEN RIB
TO HIS BREASTBONE—WITHOUT THE AID OF AN ANAESTHETIC.
THE SCREWS ARE ONE INCH IN LENGTH—HE TAKES THEM OUT DAILY TO STERILIZE THEM

Charmed Lives

Many people have defied chance over the years, cheating death or fate. For some, it delayed the inevitable. For others, they're happy to be alive:

* Henry Ziegland of Texas jilted his sweetheart in 1893 and was shot at by her brother. But the bullet only grazed his cheek and buried itself in a tree. Twenty years later, Ziegland tried to cut down the tree but decided to use dynamite. The blast dislodged the bullet and sent it spinning into Ziegland's head, killing him on the spot.

* All the births on Jefferson Avenue in El Paso, Texas, were on the south side of the street from 1922 to 1938 and all the deaths were on the north side.

* Twins Constantine and Caesar Faucher were born in France in 1759. They became lieutenants in the army on the same day, captains on the same day, colonels on the same day, were wounded in the same manner in the same battle, were decorated together, made brigadier generals together and were tried and sentenced to death for treason together. They were reprieved at the last moment together, retried, found guilty together and again sentenced to death on the same day. They were identical except for the flowers they wore in their buttonholes on the way to the guillotine.

* Sgt. Joseph Charles was in a foxhole in New Guinea when a mail boy called him out for a letter. He crawled out and a Japanese plane came overhead and dropped a bomb that totally destroyed the foxhole he'd just left. The letter was from his mother.

LUIS de CAMOENS
(1524-1580)

THE PORTUGUESE POET SHIPWRECKED IN 1561, LOST ALL HIS POSSESSIONS EXCEPT HIS MANUSCRIPT OF "LUSIAD," THE *POEM THAT MADE HIM FAMOUS*

Odd Ones

Some medical cases are incredible. But beyond the miracle cures and the chance discoveries are the following stories:

* Dick Winslow appeared at a Los Angeles hospital complaining about a pain in his throat. Doctors found a Mickey Mouse watch lodged in his esophagus.

* While taping *The Dick Cavett Show*, nutrition expert J. I. Rodale said that he was so healthy he planned to live for a long time. He then dropped dead on the spot.

* Dr. Christiaan Barnard, the famous South African heart transplant surgeon, was once offered $250,000 to perform a human head transplant.

* The U.S. Center for Disease Control says dog bites are overtaking gonorrhea as the most common affliction in America.

* Penelope Van Princis was attacked by Indians in 1620 while traveling in America after a trip from Amsterdam. Her skull was fractured and a spear was shoved through her body. She was left in a hollow tree for seven days but survived to live another 92 years and have 502 descendants.

Sir James Harrington (1511–1591) of Exton, England, married off his 18 children so well that within a century of his death his direct descendants were distributed over a large segment of English nobility, and included 8 dukes, 3 marqueses, 27 viscounts, 36 barons and 70 earls. Believe It or Not!

The Cells That Would Not Die

The story of Henrietta Lacks defies medical chance and is still baffling scientists today.

One winter day in 1951, the 31-year-old Lacks was admitted to Johns Hopkins University medical clinic in Baltimore. Doctors found a strange purple lesion about an inch in diameter within her cervix.

They cut out a section and determined it to be malignant. Henrietta Lacks had cancer.

Days later, they bombarded Henrietta with radiation but the tumor proved invulnerable and she died eight months later. But part of her is still alive!

Doctors had removed a few cells from the tumor and found to their surprise that they were thriving, doubling their number every 24 hours. Previous tests under artificial conditions had failed to produce strong cancer cells.

Researchers everywhere involved in a cure for cancer still seek Henrietta's cells, the first human cells available for cancer experimentation. They are called HeLa.

Remedies

Feel sick? Forget the drugstore. History is full of curious remedies and most involve animal parts.

* Crocodile scales sauteed in butter are supposed to cure toothaches and boils.

* A piece of a tortoise tied to the head can prevent malaria.

* Bones from monkeys boiled for 10 days are reputed to cure rheumatism.

* Powdered porcupine brain has been used as a burn remedy.

* Powdered deer antlers are sometimes used to restore virility.

* Rhino horn is an aphrodisiac; its urine is said to restore youth, and the hoof cures fevers.

* Tigers do everything. The fur cures fevers; the brain, apathy (and acne when mixed with oil); the tail, skin ailments; and the whiskers, when mixed in a potion, will protect you from bullets.

HAVOC OF A TOOTHACHE

HALBERT DeFREEST - BUFFALO, N.Y.

SUFFERING THE AGONY OF AN ACHING TOOTH
ACCIDENTALLY STRUCK A POWER POLE WHICH PULLED DOWN **11** OTHER POLES
CAUSING STREET AND HOUSE LIGHTS TO GO OUT-HALTING THE WORK OF **4000** MEN IN **8** FACTORIES
AND PRODUCED A TANGLE THAT REQUIRED **200** LINEMEN **12** HOURS TO STRAIGHTEN OUT

PART SEVEN:

The Return of the Past
The Changing Tides of History

G reat empires rise—and crumble into dust. Kingdoms are won and lost. From the first moment man started to walk along the highway of history his footsteps have been dogged—by the shadowy, menacing figure of Chance. Walk with us through the pages of history and meet the civilizations and conquerers that the Fates have smiled upon—and those they have brushed aside with a flick of their hand. Read how some men are born great, some achieve greatness—but how many have greatness thrust upon them by sheerest chance. There's:

* The random shot from a bow that changed the course of a nation.
* The accidental discovery of America.
* The sudden shift in the wind that saved England from invasion.
* The shoal of fish that cost Mark Anthony an empire.
* The incredible mistake that enabled Germany to conquer France.
* The unbelievable parallels between the deaths of Lincoln and Kennedy.

And much, much more. The history of man truly is one of bizarre coincidences, twists of fate and baffling happenstance. And it's all true!

The entry of Alexander the Great into Babylon.

Footprints of time

A pre-historic mammoth hunt.

Footprints of Time

How long has mankind walked upon the earth? Most historians agree that man didn't learn to stand upright until a million years ago and only learned to use tools about 50,000 B.C. Yet a chance discovery may force us to rethink these theories.

William J. Meister, excavating near Delta, Utah, in 1968, broke open a rock fossil about 300 million years old—and found what appears to be a footprint of a man, wearing a shoe! And the foot had crushed a trilobite back in prehistoric times—hundreds of millions of years ago!

Man and the Dinosaurs

The dinosaurs were wiped off the earth suddenly millions of years ago and man appeared on the scene somewhat later, right? Well, maybe. Consider the mysterious evidence of the sculptures of Acambaro, Mexico.

Waldermar Julsrud just happened to notice some sculptures washed out of a mud-bank there in 1945 and asked a local stonemason to dig there and bring him some samples. Since then more than 30,000 sculptures have been found there, carbon-dated to 1600 B.C. They depict dinosaurs, early mammals and what look like dragons —all supposedly long-vanished from the earth.

But what's even more amazing is that they show human beings alongside these supposedly "extinct" and mythical creatures! Did man actually live with the dinosaurs? And did dragons really exist? We may never know—but fate has certainly unearthed some fascinating clues!

Some Giant Steps...

Giants lived on earth many years ago and fought with man, according to the world's mythologies. And it's there in black and white in the Bible too. David, for example, fought with the giant Goliath and felled him with a slingshot. Most modern experts scoff at the idea of races of giants, however. But some recent unexpected discoveries may yet prove that mythology is more than just a load of tall stories:

* A giant footprint, 14½ inches long, was found at Parkersburg on the Ohio River, the *American Anthropologist* for 1896 reported. And it was wearing a shoe!

* Albert E. Knapp found another fossilized shoe-print from the Triassic period in Fisher Canyon, Nevada, complete with hand-stitched leather mocassins!

* Troy Johnson, an engineer, made plaster casts of giant footprints he found near Tulsa, Oklahoma, in 1969.

What's more, there's a strong possibility that giants are still around today. They crop up in tribal legends throughout the world and have been sighted often in modern times. Several expeditions to Mount Everest and elsewhere in the Himalayas have reported running into the terrifying abominable snowman or yeti; there are numerous accounts of close encounters with the sasquatch, or Big Foot, in the north-western forests of the United States and in British Columbia.

For example:

* Big Foot has been caught on film—by Roger Patterson in 1967. He shot 30 feet of film of a giant, hairy, humanlike creature striding through the woods of northern California. It was about 6 to 6½ feet tall, had extremely long strides that could not be made by a human, and vanished quickly. There's absolutely no evidence that the film was faked. It's an astonishing, tantalizingly brief glimpse of a creature that may date back millions of years!

* A report of a human making contact with a sasquatch came in November 1981, when 13-year-old farmer's daughter Tina Barone was confronted with a beast with

glaring red eyes that towered over her. "Its fur was about one inch thick and all matted and dirty," said Tina.

* The existence of the sasquatch was officially recognized by the U.S. Corps of Engineers in 1975. In its *Washington Environmental Atlas*, it said: "Reported to feed on vegetation and some meat, the sasquatch is covered with long hair, except for the face and hands, and has a distinctly humanlike form." There was a sketch of a furry upright beast, which the corps said is agile and strong, up to 12 feet tall, weighs more than half a ton and takes strides of up to six feet.

The Great Flood

Many ancient legends tell of a Great Flood that occurred in prehistoric times and killed most of mankind. The story of Noah in the Bible reveals that a few survivors sailed atop the waters in an ark, survived the stormy waters and preserved the race. But what caused the Great Flood? Was it the wrath of God or a chance act of nature?

Emmanuel Velikovsky, in his book *Worlds in Collision*, theorizes that the Great Flood was caused by a near collision among the planets:

* Venus, he says, only became a planet relatively recently. Previously it was one of Saturn's moons and was torn loose by a disturbance in the electromagnetic balance of the solar system. Racing past Jupiter, Venus became a comet, nearly colliding with Mars and Earth, and causing the Great Flood as it passed near Earth's orbit.

* The parting of the waters of the Red Sea and many other ancient catastrophes and miracles were also caused by the near miss. The earth's magnetic poles reversed, the Ice Ages began and the earth "stood still" at the Battle of Jericho!

Many experts are skeptical of Velikovsky's work. But he did predict the surface temperature of Venus with uncanny accuracy!

Gods on Earth: Von Daniken's Views

Civilizations rise and fall, but the history of man is one of steady progress from the cave man to the astronaut. But how then do you explain some of the marvelous artifacts of the ancient world? Things like the pyramids of Egypt, seemingly impossible creations for a people with a primitive technology. The controversial Erich Von Daniken believes mankind was given a nudge forward by visitors from outer space who happened to drop in on Earth. Among his suggestions:

* Most world cultures, including the Hindu epics of India, the Mesopotamian myths and the Central American legends, say man's civilization started with visitors from the stars.

* The pictures of the early "gods" who brought these arts and sciences all show creatures in costumes that resemble space suits and helmets, many of them descending in "chariots" that look like reentry capsules.

* Many early inventions seem to foreshadow modern technology. The Ark of the Hebrews, he says, is like an early radio set with its listening apparatus and its deadly electric currents. The Hindu "flying machines" are like airplanes. The force that destroyed Sodom and Gomorrah could have been an early atomic blast.

Most scholars say the weird and wonderful artifacts that Von Daniken cites had religious or artistic purposes. But there's always the chance Von Daniken is right.

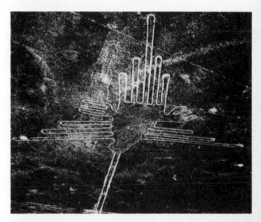

Unexplained markings from a mysterious ancient civilization in the Nazca Valley, Peru.

Ancient Technology: Some Modern Finds

Many ancient objects have been dismissed as merely ornamental until reassessed by modern experts:

* A primitive spark plug, discovered at Owens Lake, California, in 1961, was compared with modern spark plugs by a mechanic and found to be accurate in every respect. The artifact is thought to be 500,000 years old!

* A primitive battery was found by German archeologist Wilhelm Koenig near Baghdad in 1938. A modern replica, built by an electronics expert at General Electric, actually worked! There's evidence that the ancient Babylonians had electroplating, and an ancient Sanskrit text, long thought erroneous, gives clear instructions for making batteries.

* An ancient computer, found near the island of Antikythera in 1900, was analyzed recently by Dr. Derek de Solla Price of Princeton University. It dates back to 65 B.C. or earlier and can be used to work out the motion of the sun, moon and planets in the solar system. Price says the discovery is like finding a jet plane in the tomb of King Tut!

The Mystery of the Silent Stones

Modern archeologists have unearthed ancient cities and monuments—only to be baffled at the incredible sophistication of what they found. Here are some examples:

* Machu Picchu, a massive fortress built high atop the cliffs of the Andes in Peru, has stone blocks so heavy they couldn't be moved by modern machinery. It also has a harbor, thousands of feet above any modern sea.

* The Temple of Jupiter at Baalbek, Lebanon, is made of stone blocks weighing from 750 to 1,000 tons. Again, how they could be moved by a primitive people is a mystery.

* A cut-stone block at Sacsahuaman, Peru, weighs 2,000 tons and has been left upside down, ready to be moved!

* Red porphyry blocks at Ollantaytambo in Peru must have been carried thousands of miles over rivers and through jungles, then hoisted to the top of a 1,500-foot cliff—by people who didn't know about the wheel!

* Deserted Pacific islands contain many stone monuments, including a mammoth stone tomb weighing 170 tons in the Friendly Islands and a 40-ton monolith on Manua Levu in the Fiji Islands.

The Skull of Doom

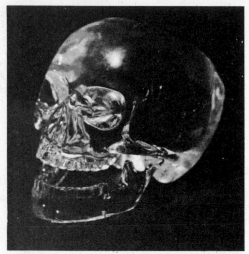

A mysterious crystal skull was unearthed by Anna le Guillon in 1927 at an excavation at Lubantuum in British Honduras. It weighs 11 pounds 5 ounces, and is carved of pure quartz in such a way that light that is shone from underneath glows through the eyes!

Anna's adopted father, adventurer "Mike" Mitchell-Hedges, said the skull is "the embodiment of all evil."

Authorities say it took 150 years of polishing with primitive technology to achieve its present perfection but disagree on whether it is of European or Central American origin. The lifelike skull has a hinged lower jaw which works perfectly!

Mapped from on High

An ancient map, used by the Turkish admiral Piri Reis and said to have been drawn from much older maps, has survived from the days of Columbus. For years it was thought to be inaccurate and meaningless because the continents on it were so askew. But modern aerial charts show it to be amazingly accurate—except for the outline of Antarctica. Then it was compared with sonar depth soundings, and it matched Antarctica perfectly—without the ice-sheet!

Christ's Prophets

The Old Testament prophecies of the birth of Christ are well known. But the coming of a savior was also hinted at by such people as the Greek philosopher Plato and the Roman poet Vergil.

* The philosopher-king in Plato's works, written around 500 B.C., is said to be destined to one of two fates. He would be declared absolute ruler or be crucified as a criminal.
* In Vergil's *Eclogues*, written between 42 and 37 B.C., the birth of an infant savior is predicted, though the poet had never heard of Judea or of Christ's ministry!

Omens of Disaster

The Fates warned the emperor Julius Caesar of his impending death—but he chose to ignore them.

The night before his assassination in 44 B.C., his wife dreamed that a statue ran blood and she warned Caesar not to go to the Senate that day, the Ides of March. The warning was repeated by a soothsayer but Caesar brushed them aside as nonsense.

He was stabbed to death by Brutus and other senators who feared he was about to set himself up as a dictator. Civil war followed and the winner, Augustus, founded a line that ruled Rome for 400 years!

Taking No Chances: The Birth of Alexander

Philip of Macedon married Olympias of Egypt, whose father, Neatanebus, happened to be an astrologer. He cast a horoscope for Olympias, telling her, "Your child will become master of the world." He even persuaded her to postpone the birth until exactly the right hour! Sure enough, a world leader was born—Alexander the Great, who is said to have wept at the age of 23 because he had no more worlds left to conquer!

End of a Dream: The Death of Alexander

Alexander died tragically early at the age of 33—the result of swamp fever and a drinking bout. He was conqueror of most of the known world, but his empire quickly fell apart as his generals fought for power. But consider what might have been if fate had not deprived the world of this outstanding man of action and ideas. Alexander believed in:

* Social equality and intermarriage among races. He held mass marriages between his troops and his subject people and insisted all native beliefs and religions be honored.
* One language. Greek became the language of the civilized world, and the cities he founded, like Alexandria in Egypt, became centers of culture and learning for thousands of years. Greek medicine, laws, science and poetry all flourished.
* One government. Building on Persian practices and adding the Greeks' respect for individual rights, Alexander founded a forerunner of the United Nations, under one ruler.

How Anthony Was Suckered

According to Shakespeare, Anthony deserted his fleet at the Battle of Actium in 31 B.C. to run after Cleopatra, who was heading home to Egypt. But it wasn't love that cost Anthony the crucial battle—it was a plague of sucker fish! They fouled the oars of his lead ship and made it impossible to go on. Anthony left the ship and his dispirited sailors retreated, leaving the victory—and the empire—to Octavius Caesar!

Epilepsy by Chance

Christianity and Islam are two of the world's major religions today. But they may owe much of their success to—attacks of epilepsy.
*Paul was a Pharisee who persecuted Christians in Tarsus. He had a vision of Christ on the road to Damascus—probably due to an epileptic attack. He became the chief ambassador for the new faith, traveling all over the Mediterranean to spread the word.
*Muhammed was a recluse, subject to epileptic fits, when he had a vision of the angel Gabriel in 610, telling him of the word of God. He took the message to the Arab communities of Mecca and Medina. And by the time of his death, the faith was well established in the Near East.

Muhammed had no sons and no obvious successor. A major split occurred over the status of subsequent caliphs or successors. The minority Shiite sect believes Muhammed's son-in-law Ali to be the first true successor. The majority Sunni sect holds that Ali's three predecessors were legitimate caliphs.

The Sign of Victory

A vision in battle saved Christianity from possible destruction.

In the midst of a crucial battle against the barbarians in 312 A.D., Constantine suddenly had a vision of a cross in the sky, complete with the words "By this sign you will conquer."

He won the battle and adopted the new faith, which was still struggling to survive in the Roman Empire. Christianity became the official religion of the Roman world and its future was assured.

Hallelujah!

A heathen army was attacking the forces of the bishop of Auxerre at Mold in Wales on Easter Sunday in 430 A.D. The bishop yelled the battle cry of "Hallelujah!," his troops took up the cry—and the enemy fled in confusion!

How Christianity Came to France

A vow made in the heat of battle was responsible for France becoming a Christian nation!

King Clovis, who was married to Clotilde, a Christian, was about to be overwhelmed at the Battle of Tolbiac in 496 when he exclaimed, "God of Clotilde, if you grant me to win this battle, I'll become a Christian like her." He won and France turned to Christianity!

Their Loss Was His Gain

The Byzantine emperor Constantine V and his army of 80,000 men lined up against a Bulgarian army of 80,000 in 743 A.D. Constantine gained an astonishing victory. Not one man from his army was killed or even wounded. The Bulgarians lost 20,000 men!

THE **MAN** WHO CHANGED THE FAITH OF MILLIONS OF PEOPLE ON A WHIM
RAINCHAN SHAH, THE TIBETAN WHO CONQUERED KASHMIR IN 1325, BECAUSE HE WAS A LOW-CASTE HINDU DECIDED ONE DAY TO ADOPT THE RELIGION OF THE *FIRST MAN HE MET THE FOLLOWING MORNING*—HE MET BULBUL SHAH, A MOHAMMEDAN – **AND AS A RESULT ISLAM BECAME THE OFFICIAL RELIGION OF KASHMIR**

1066 and All That

A random shot from a bow changed the course of a nation back in 1066.

Harold Godwinson, the Anglo-Saxon ruler of England, was holding off the invading forces of William the Conqueror when William turned to trickery. He ordered his cavalry to retreat, drawing many of the English foot soldiers after them. The cavalry then wheeled and charged, slaying a large number of the foe. He then ordered his archers to shoot up in the air, instead of directly at the enemy. Harold looked up—and was fatally wounded by an arrow in the eye.

The despairing Anglo-Saxons rallied around Harold's body but the battle was lost. And William promptly took control of the island, changing the legal language to French, granting large tracts of land to his followers and disinheriting many of the great Saxon families. For the last time in its history, a successful invasion of England had been mounted.

The Prophetic Jest

A young priest was injured and robbed by thugs in 1321 and nursed back to health by the prior of the Monastery of Thuet.

"When will I ever be able to repay you?" the priest asked. "When you are pope," the prior replied.

Twenty-one years later, the young priest was crowned Pope Clement VI and he appointed the prior Archbishop of Toulouse.

A Royal Dream

A humble soldier named Romanus dreamed in 1065 that he was sitting beside the Byzantine empress Eudokia. He told the dream to a guard and was arrested and sentenced to death, but the empress pardoned him! Two years later, when Emperor Constantine X died, she married the soldier—and he became Emperor Romanus IV!

The Hard Times of Old England

The legend of Robin Hood and His Merry Men, robbing the rich and giving to the poor, is probably based on the exploits of several men who suffered under the rule of the regent, John. And the hard times and civil disorder all came about because of a navigational error.

John's brother, Richard I, was setting off for the Crusades in the Middle East when one of his navigators misread a map. He landed in Algeciras and was promptly imprisoned. John took over—and the nation suffered until Richard the Lionhearted was finally ransomed.

Legend has it that Richard was finally found by chance. His own balladeer searched for him all over Europe and found him when the second verse of a song Richard had composed came floating out of a jail in answer to the first verse which the minstrel had sung!

A Long, Long War

The Hundred Years' War between France and England went on long enough—but in fact it hasn't officially ended yet! There was only a truce called between the two sides—and many were made and broken during the course of the war between 1337 and 1453.

The English should never have stood a chance. France at that time was rich and prosperous, while England was poor and underpopulated. France had a population of about 11 million—three times that of England. Mounted on giant war-horses, the magnificent French cavalry was a daunting force, the equivalent of today's heavy armor. But the English longbow—which could shoot two or even three arrows for every crossbow quarrel —brought the English some amazing victories:

* At the Battle of Crecy in 1346, the English had about 11,000 men—but were outnumbered nearly three to one by the French. A line of Genoese crossbowmen was cut down by a storm of arrows. A hopelessly disorganized cavalry charge followed and was again met with a murderous rain of arrows. The French charged 15 times, and the battle went on well into the night. The English, who had lost under 100 men, didn't realize the slaughter they had inflicted. When day dawned, they saw the field littered with French bodies. An estimated 10,000 had died.

THE MONARCH WHO VALUED HIS HONOR ABOVE HIS LIFE !

KING JOHN II (1319-1364) of FRANCE, CAPTURED BY THE BRITISH AT THE BATTLE OF POITIERS AND THEN FREED BY LEAVING HIS SON, THE DUKE OF ANJOU, AS A HOSTAGE, UPON LEARNING THAT HIS SON HAD BROKEN HIS PAROLE BY ESCAPING, RETURNED VOLUNTARILY TO LONDON--WHERE HE DIED

6,000. They were starving, and weakened by dysentery and a miserable march through the cold. The French, with perhaps 50,000 men-at-arms, were bustling and confident. The muddy ground broke up the French charges. The English longbowmen sent clouds of arrows at the French. In less than four hours the English were triumphant. The French lost about 10,000 men, the English 300, including Henry's cousin, the fat duke of York. He had fallen over and been suffocated by bodies falling on top of him!

Laughter Was the Winner

The Battle of Buironfosse in France, in 1339, is usually thought of as one of the opening encounters of the Hundred Years' War. In fact it never took place—and all because of a frightened rabbit which dashed between the lines of the two opposing armies. The sight was so hilarious that the soldiers on both sides roared with laughter—and withdrew without exchanging a blow!

Battle-scarred

Mallet Eustache, governor of the province of Guise, France, was wounded in each of the 122 battles he fought—but died peacefully in his sleep in 1349!

Fooled, by George

Daniel de Bouchet, a Burgundy knight, had led eight unsuccessful attacks on the English holding the fortress of Braine-LeCompte in 1423. So, on his ninth try, he decided to see if he could fool the English into surrendering —by dressing up as St. George, the patron saint of England. It worked! The English thought he was a vision and gave up without a struggle.

Thrown Together

Francesco delle Barche invented a giant catapult that could hurl a 3,000-pound missile. During the siege of Zara, Dalmatia, in 1346, he became entangled in the catapult and was hurled into the town. Unbeknown to him, his wife was in the town. He landed on her—and both were killed!

* The English didn't even want to fight the Battle of Poitiers in 1356. The Black Prince's men were retreating, exhausted and laden with plunder. In fact the two armies didn't know how near they were to each other until the English advance guard accidentally collided with the rear of the French army!

The Black Prince tried to negotiate, offering to return all the town, castles and prisoners he'd captured and to swear not to take up arms against the French for seven years. The French king John, anxious to revenge Crecy, wanted nothing less than unconditional surrender. He had 16,000 to 20,000 men, while the Black Prince had only 6,000 to 7,000, mainly foot soldiers and archers. Despite favorable terrain—covered with undergrowth, hedges and patches of marsh— the English were in desperate straits, until the duke of Orleans (King John's brother) inexplicably decided not to attack and marched off the field! The fight continued and the English won the day!

* At Agincourt in 1415, the English army under King Henry V numbered less than

America: Discovered by Mistake

BERTRAND.

America was officially discovered in 1492 when Christopher Columbus set out with financial aid from the king and queen of Spain. But what the history books fail to point out is that it was all a complete accident!

* Columbus was, in fact, trying to discover a shortcut to the Orient, so he could bring the fabulous riches of the East back to his sponsors.

* Arriving in the Caribbean, he wrote in his ship's log that he was in India.

* When he reached Cuba, he proclaimed that the ships of "the mighty Khan" of China went back and forth from there in 10 days!

* It was only on his third voyage that he began to suspect he may have chanced upon "a very large continent which until now has remained unknown."

* Columbus died in 1506, having brought neither gold nor spices from the East. And America was named after someone else— Amerigo Vespucci!

* Spain, however, came out a winner. By investing about $6,000 in Columbus's first voyage, the monarchs sparked a boom in exploration which eventually netted Spain more than $1,750,000—in just over 100 years!

His Satanic Majesty: A Modern Invention?

The Devil, as we know him today, may be a chance spin-off of the witch trials and the Spanish Inquisition during the Renaissance. There are passing references to a "Satan" in the Bible, and fertility cults across Europe speak of a "horned god." Yet the modern Devil's connection with black masses and covens arose for the first time in the Middle Ages. Authors such as Dennis Wheatley suggest the Devil is the invention of a strong Christian church, to stamp out the remains of nature-worship among its parishioners!

Knights of the Garter: A Wizard Idea?

An accident at a dance in medieval England may have exposed an early coven, centered on the royal court.

When the countess of Salisbury's garter slipped at the court dance held by Edward III, he swept it up, crying, "Honi soit qui mal y pense" (Evil be to him who evil thinks). He then made the garter the foundation of a new male order of 26 knights—double the number of witches in a coven. Some writers suspect that the garter may have denoted a queen of the witches and the king was establishing a new male coven, with himself as head of the Old Religion!

Jolly Sporting

Louis, duke of Longueville, was captured by King Henry VIII in 1513. But the French military commander was unable to raise his ransom. So the king allowed him to win the money by wagers at croquet!

Flour Power

General de Bellay saved the besieged and starving garrison of Landrecies in France in 1543—by tying bags of flour onto 600 horses, equipping them with lances, swords and helmets, and leading them on a dark and foggy night through 40,000 soldiers. They went through unchallenged and the city was saved.

The Horse That Solved a Murder

Jean Poltrot, a Huguenot, assassinated the duke of Guise, leader of the royalists in France's religious war, on February 19, 1563. He rode off on a horse that he'd bought from a thief and, after riding hard all night, discovered he was back at the scene of the crime! The horse had been stolen from the duke's stable and when it came home, Poltrot was captured and executed!

The Heavens Speak

The leader of the Protestant Reformation in Germany turned to religion because of a storm.

Martin Luther (1483–1546) was walking on the road near Erfurt during a storm when a flash of lightning knocked him down. He was filled with a fear of death and a consciousness of sin.

He renounced the world and entered the monastery of the Augustinian Eremites.

THE **THUNDERBOLT** THAT SAVED A WOMAN FROM BIGAMY!

ELIZABETH SYDENHAM, *WIFE OF SIR FRANCIS DRAKE*, ADVISED THAT HER HUSBAND HAD BEEN KILLED BY THE SPANIARDS, WAS ON HER WAY TO CHURCH TO MARRY ANOTHER MAN WHEN A BOLT OF LIGHTNING STRUCK THE GROUND AT HER VERY FEET! ELIZABETH INTERPRETED IT AS A SIGN THAT HER HUSBAND WAS STILL ALIVE AND CALLED OFF THE WEDDING

LATER SIR FRANCIS RETURNED FROM HIS NAVAL EXPEDITION—ALIVE AND WELL

The Winds of Fate

England's emergence as a major world power dates from 1588 and the defeat of the Spanish Armada. Not only did the English have brilliant captains such as Sir Francis Drake on their side, they also had Nature!

The fast, maneuverable English ships had already inflicted severe damage when the wind suddenly changed, forcing the Spanish fleet to head north round Scotland to Ireland and back home by the Atlantic. Some 130 ships had set sail from Spain; fewer than half reached home and only a third of their men survived! Thirst, hunger and the violent sea carried off the rest. Many ships put in for food and water in Ireland, only to be caught in a vicious hurricane that battered them to pieces. Most of their crews were drowned, but over 1,000 fought their way through the surf—only to be stripped naked and slaughtered by the Irish and by bands of English soldiers!

Final Words

King Charles I of England, a lifelong stammerer, uttered only one sentence without stuttering—his farewell message on January 30, 1649, just before he was beheaded!

Humble Beginnings

Sir George Rooke (1650–1709) was sent to sea as a boy by his father for petty thievery. He went on to fame as the British admiral who captured Gibraltar.

The Cat and the Queen

The marquise de Maintenon, second wife of Louis XIV of France, lived to become queen—thanks to a cat.

Francoise d'Aubigne, daughter of the French governor of the island of Marie Gallant, was just 3 when she was pronounced dead at sea and her body sewn up in a sack to be thrown overboard. Her pet kitten had crawled inside the sack and began meowing during the funeral service. Knowing cats shun corpses, the captain had the sack ripped open. The little girl was still alive, went on to marry the Sun King, and died in 1719 at the ripe old age of 84!

WHEN THE PILGRIMS LANDED IN 1620 THEY WERE GREETED BY AN INDIAN WHO SPOKE ENGLISH!

IN 1614 CAPT. THOMAS HUNT RAIDED THE COAST OF NEW ENGLAND AND CARRIED OFF 20 INDIANS WHOM HE SOLD AS SLAVES IN SPAIN - EXCEPT ONE WHO WAS SMUGGLED BACK TO NEWFOUNDLAND AND FINALLY RESTORED TO THE MASSACHUSETTS TRIBE BEFORE THE PILGRIMS ARRIVED.

A Flaming Oversight

The Great Fire of London in 1666 turned London into an inferno—because nobody thought it was serious enough to go to the bother of putting it out!

It all started in Pudding Lane, in the shop of John Farynor, baker to Charles II, and spread to a pile of hay outside. Crowds gathered to watch, but nobody was particularly concerned. When the lord mayor was informed, he shrugged it off with the words "A woman might piss it out!"

When the fire reached the Thames River, barges loaded with timber, oil and coal burst into flame. Fanned by a strong east wind, it spread to the city's financial area. Shops on London Bridge were gutted and St. Paul's Cathedral was destroyed—but only eight people died because there was plenty of time to flee. The smoke was so dense, says Samuel Pepys in his diary, that it "darkened the sun at midday."

But the Great Fire brought some accidental benefits. The filthy, overcrowded and disease-ridden London slums were wiped out, and the last of the areas that had been ravaged by the Great Plague the previous year were cleansed—by fire. And the genius of architect Christopher Wren flourished in the massive rebuilding program that followed, setting the pattern for the London of today.

The Song That Saved 400 Lives

Four hundred Welshmen invading St. Malo, France, on September 6, 1759, sang an ancient Breton song as they advanced. A Breton army was waiting to ambush them. But when its soldiers heard the song, dating back to the days when the Welsh and Bretons belonged to one nation, they got up and joined in the song.

Both sides ignored their officers' orders and the invaders reembarked—without a single shot having been fired on either side.

Divine Intervention

Montezuma II, Aztec ruler of Mexico.

The powerful Aztec empire in Mexico was subdued by the Spanish (1519–1521) under Cortes—by a freak accident. Cortes was advancing on the Aztec capital with 100 men and a few cannon when many of the enemy fell to their knees and started worshiping his lieutenant, Pedro de Alvarado!

Alvarado was blond and bearded, a rarity among the swarthy conquistadores, and the Aztecs believed he was their god, Quetzalcoatl, who had departed centuries before, vowing to return.

The surprise and religious awe generated by Cortes's lieutenant tipped the balance in favor of the Spanish; they took the town with ease and quickly conquered the entire countryside, shipping back gold to Spain in unbelievable quantities, making the nation rich beyond its wildest dreams!

The Welsh Hat Trick

A French army that landed at Pembroke, Wales, in 1797, surrendered in panic when the Welsh made them believe they were outnumbered. How? By getting squads of women to march up and down the distant hills, wearing tall black beaver hats!

The Fall of the Incas

The all-conquering Spanish also humbled the mighty Incas of Peru. But here again, they were aided by a huge dollop of luck!

The Incas were rich, powerful and efficient. Amazingly skilled in agriculture, they had domesticated hundreds of plants—including the potato which was later introduced to Europe. They had built hundreds of miles of roads across mountains and through dense jungles, and communications were so good that the Inca ruler was updated on major events in a matter of days. They had a huge standing army, and quashed rebellions with awesome ease.

That is, until the arrival of Francisco Pizarro and his far-from-overwhelming number of troops. They had the supreme advantage of gunpowder. They also had the Fates on their side.

It just so happened that the previous Inca ruler, Huayna Capac, had died in 1525 without leaving a clear successor. His two sons had just fought it out to the finish when Pizarro arrived. The winner, Atahualpa, had not yet been crowned—he was on his way to his coronation from the local baths when the Spanish captured him!

He desperately tried to buy his release with a roomful of silver but the Spanish killed him and wiped out the last resistance of the dispirited Incas. Had he been crowned and been able to defend the kingdom from the Incas's mountain fortresses, the war might have gone very differently!

Transport of Delight: The Spanish Mane

The American Indians dragged their possessions around on dog sledges for many centuries. But all that changed with the arrival of the Spanish. Horses escaped from their corrals and ran wild in the American southwest—and the Indians were quick to capture them and use them as a means of transport.

By an odd quirk of fate, prehistoric horses had roamed the plains there many centuries before—but had become extinct!

Arrow of Fate

Sir Henry Morgan was having no luck in 1665 trying to storm the Spanish fortress of San Lorenzo, which contained the wealth of Panama. Then fate lent a hand.

One of his soldiers was hit in the shoulder by an arrow. He pulled it out, put a wad of cotton around the point and fired it from his musket. It caught fire, and set a thatched roof alight. The flames spread to the powder magazine which exploded, killing 363 of the 400 soldiers inside the fortress!

The Northwest Passage: Breaking the Ice

Since the days of Columbus, man had dreamed of the day he would find a northern route to the rich and mysterious East. But it was only in the 20th century that icebreakers finally forced their way through the wintry waters of the Arctic and made the dream of centuries a reality.

But during their attempts to find the passage, the French and English probed the eastern waterways of Canada and the U.S., mapping the St. Lawrence, the Great Lakes and Hudson Bay, laying the groundwork for the settlement of Canada and the northern United States.

The Tax Man Cometh

Samuel Adams, who organized the Boston Tea Party in 1773 to protest British taxes on the colonists, had been the Boston tax-collector in 1764!

A Real Gentleman

Maj. Patrick Ferguson's sense of honor changed the course of history!

Ferguson, leader of a corps of British sharpshooters, could have shot Gen. George Washington near Brandywine Creek on September 7, 1776—but considered it un-unsporting to shoot a man in the back.

Walking Wounded

A soldier in the American Revolution had a 98 percent chance of surviving in battle—but only a 75 percent chance of surviving a stay in an army hospital.

Connecticut Yankees

The Order of the Purple Heart was created by George Washington in 1782 to commemorate wounds received in battle. But it was awarded just three times in the next 150 years—each time to men from Connecticut!

Spared by Fate

British soldier and statesman Robert Clive (1725–1774) tried to commit suicide twice when he was young. But he was spared by chance—when his pistol misfired. He went on to fame and fortune but killed himself in his third attempt at 49.

Trapped—by a Donkey!

Louis XVI of France fled Paris in 1791 with a republican mob hot on his heels. And he might well have got away—but for a stubborn donkey.

The king's coach had reached the gates of Vincennes, but his sole escape route was blocked by a farmer's cart and donkey.

Frantic efforts to persuade the donkey to move were all in vain—the delay allowed the mob time to catch up, seize the king and haul him back into Paris.

A year later Louis and his queen, Marie "Let them eat cake" Antoinette, were tried for treason and sent to the guillotine!

To Arms, Citizens

Claude Joseph Rouget de Lisle, who composed the "Marseillaise"—the national anthem of the French Revolution—was a royalist who only narrowly escaped the guillotine!

Making His Fortune

Fortune-teller Marie-Anne Lenormand was consulted in 1795 by an officer in the French army who was frustrated by his prospects for promotion. She persuaded him to stay in the army and predicted he would rise to high rank. The officer's name? Napoleon Buonaparte!

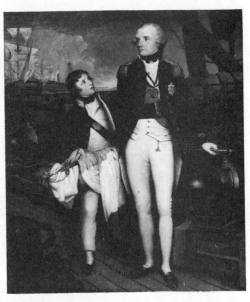

Kismet, Hardy

Lord Nelson, fascinated by West's painting "The Death of Wolfe" and flattered by the artist's promise to paint him in the same manner, remarked, "Then I hope to die in the next battle." He did—at Trafalgar—and the artist kept his word.

A Nasty Cough

Napoleon once coughed—and sentenced 1,200 people to death.

He was a general in the Middle East in 1799 and had just decided to release 1,200 Turkish prisoners.

Then came the fatal cough. Napoleon exclaimed, "Ma sacrée toux" (My confounded cough). His officers thought he said "Massacrez tous" (Kill them all). And they executed the prisoners!

Abigail Willing, could have become a future queen, but she rejected a proposal from Louis Philippe, the Duke of Orleans, while he was in exile in the U.S.A. Her father, a prominent U.S. banker, told Louis Philippe: "If you have no chance to the throne of France you are no match for my daughter." Louis Philippe eventually became king of France and ruled for 18 years. Believe It or Not!

Retreat from Moscow

It was the dread hand of disease more than the icy fingers of winter that forced Napoleon's men to retreat from Moscow in 1812, modern military historians believe.

The flu killed many of his army; typhus, diphtheria, dysentery and colic took their toll of hundreds more—even before they got to Moscow.

And the winter of 1812, depicted in paintings as a snow-swept, blizzard-filled Russian nightmare of cold, was in fact quite warm! Frost did not arrive until October 27, and the famous crossing of the Berezina River during the retreat was dangerous because the river wasn't frozen enough.

Waterloo: A Pain in the Stomach

Stomach ulcers probably cost Napoleon the Battle of Waterloo in 1815!

The duke of Wellington and the English were caught unawares when Napoleon appeared before them. But they had time to muster their forces when Napoleon decided to delay his breakfast because of stomach troubles brought on by ulcers!

Patriotic Inventor

Rev. Alexander John Forsyth, British inventor of the percussion gunlock, which revolutionized warfare, refused to sell his invention to Napoleon for £50,000. He gave it to the British government in 1808 for a mere £2,500. And the money wasn't paid until 35 years later—six months after his death!

The Nonwar War

British orders that led to the War of 1812 were repealed the day before the U.S. declared war. But the news couldn't possibly reach the Americans in time—so the orders went ahead as planned.

The Elephant Gunner

Kubadar Moll, an elephant used to transport cannon in the Battle of Lucknow, India, in 1858, took a torch from a wounded gunner and fired one of the cannons. The elephant's action held up the enemy long enough for reinforcements to arrive.

Brotherly Love

Capt. Percival Drayton, commanding the Union frigate *Pocahontas* in the American Civil War, captured the island of Hilton Head, North Carolina—overcoming a valiant defense by the Confederate general Thomas F. Drayton, his brother!

Despite the war, the brothers remained close. And when Percival died in 1865, he left Thomas $27,000 in his will!

Ahead of the Game

The Confederate general Henry Heth (1825–1888) survived the Battle of Gettysburg—because his hat was too big! He'd stuffed it with paper which deflected a bullet that hit him in the head. Unconscious for 30 hours, he recovered and lived another 25 years.

An Ill-starred Wedding

Tragedy followed tragedy on the wedding day of Princess Maria Vittoria and the duke D'Aosta, son of the king of Italy, on May 30, 1867.

The bride's wardrobe mistress hanged herself, the gamekeeper of the royal palace cut his throat and the colonel leading the procession to church dropped dead of sunstroke. The best man shot himself, the registrar had a fatal apoplectic stroke, the stationmaster was run over by the honeymoon train and the king's aide fell from his horse and was killed!

DANIEL WEBSTER

MIGHT HAVE BEEN PRESIDENT—*BUT FOR HIS PRIDE*

HE WAS OFFERED THE *VICE-PRESIDENTIAL* NOMINATION **TWICE** WITH HARRISON AND TAYLOR — *AND* **REFUSED!**
BOTH HARRISON AND TAYLOR DIED IN OFFICE

Into the Valley of Death

The story of the Charge of the Light Brigade in the Crimean War (1854–1856) is retold from generation to generation as a supreme example of raw courage and heroism. In fact it was the most mismanaged cavalry charge in history, made immortal by the stirring words of a poem published in a newspaper!

Lord Raglan ordered Lord Lucan to charge the Russian front and prevent retreating soldiers from hauling away their guns. But the message was wrongly delivered and Lucan told his 600 men to charge the Russian guns in another part of the field. Only 329 men returned from "the valley of Death." But Alfred, Lord Tennyson wrote a poem in the London *Times* glorifying the deeds of the "noble six hundred." The leader of the charge, Lord Cardigan, survived and became a hero! "It was," he said, "a mad-brained trick but no fault of mine."

"It's magnificent," said a French general after the charge, "but it's not war."

Custer's Last Stand: A Place in History

Gen. George Custer was a proud, ruthless soldier who accomplished little during his career. But an ignominious defeat assured him of his place in history!

Promoted to brigadier general during the last days of the Civil War, Custer was reduced to captain in the peacetime army. But within a few years, his swaggering had regained him his former rank. Sent to pacify the rebellious Sioux under Sitting Bull in 1876, he split up his troops—and rode right into an ambush. He was surrounded by thousands of mounted Indians under their chief, Crazy Horse. The Seventh Cavalry detachment was killed to the last man. The body of Custer proved hard to identify—he'd had his famous long blond hair cut just before the battle!

And Chief Sitting Bull had seen the battle in a vision just days before, after dancing and feasting for 18 hours. A voice had told him that the white men would die "because they had no ears" for the complaints of the Indians.

A Fatal Mistake

1. The scene of the first attempt: where the bomb was thrown before the archduke's arrival at the town hall in Sarajevo.

2. Photographed a few minutes before the assassination: the archduke and the duchess leaving the town hall.

3. The assassin Prinzip arrested after his crime.

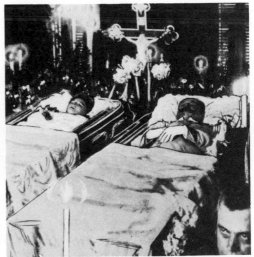

4. The lying-in-state of Franz Ferdinand and the duchess.

Phillip Prinzip was trudging miserably home for lunch. He and three other would-be assassins had missed their chance to kill Archduke Ferdinand in Sarajevo on the morning of June 28, 1914.

Suddenly, incredibly, the archduke's carriage appeared before him—the driver had made a wrong turn and was heading down the same street! Prinzip made no mistake this time—and shot the heir apparent dead. It was the shot that rang around the world. Within weeks, Austria-Hungary was at war with Serbia, touching off the "war to end all wars" that cost a staggering 18 million lives.

Umberto's Double

The Italian king Umberto I (1844–1900) had a double named Umberto Santini. Not only did they look exactly alike, they were both born in Torino on the same day. Umberto became a restaurateur the day Umberto became king. Both had wives named Margherita and sons named Vittorio and both died in Monza, Italy, on the same day. Both were shot dead—the king was assassinated and his double died in a gun accident.

Uprooted by War

The Peace Poplar, planted in Jena, Germany, in 1815 to celebrate the end of the Napoleonic wars, toppled suddenly 99 years later on August 1, 1914—the start of World War I.

Phantom Rough Riders

Teddy Roosevelt's Rough Riders, famous for their charge up San Juan Hill in the Spanish-American War, were not on horseback! The legend was fostered by imaginative journalists.

On a Wing and a Prayer

The luckiest flyer in all history must be Lieutenant Bohrle of the German air force. He was an observer in a plane cruising over the French lines at 13,000 feet in 1917 when the motor suddenly failed. As the plane plummeted Bohrle was hurled out. Then the pilot felt a bump—and there was Bohrle back in the plane, blown back by a sudden squall of wind. The pilot managed to get the plane going again and they landed safely behind their own lines!

He Did Both!

Harry S. Truman, 33rd president of the U.S., was an army captain at the front in World War I. And he once told his men, "I'd rather be right here than be president of the United States!"

Prophecy That Hung by a Thread

A statue of the Madonna on top of Albert Cathedral in France was dislodged by a German bombardment in 1915. But it dangled precariously from its thin support for three years. Frenchmen insisted that the war would end when the statue fell. German shells finally hurled the statue to the ground on November 10, 1918—and the armistice was signed the next day!

Big Bertha: A Misfire

A mix-up led to the naming of Big Bertha, the famous guns used to bombard Liege, Antwerp and Verdun in World War I. They were named in honor of Frau Bertha Von Volen, of the Krupp family. But the guns didn't come from the Krupp gunworks—they were made by the Skoda works in Austria.

2 BRITISH Q SHIPS CAPTAINED BY COMMANDER GORDON CAMPBELL IN WORLD WAR I, WERE VISITED HUNDREDS OF MILES AT SEA BY A THRUSH *WHICH FLEW INTO THE CAPTAIN'S CABIN* COMMANDER CAMPBELL'S SHIPS SANK 3 GERMAN SUBMARINES *–EACH ONE ON THE DAY FOLLOWING A VISIT BY THE THRUSH*

Flagg-wavers

One of the holders of the first number in the U.S. military draft in 1917 was a man aptly named Alden C. Flagg. His son, Alden C. Flagg, Jr., held the first number of the U.S. peacetime draft of 1940!

Here We Go Again

The U.S. troops fought their first major battle against the Germans on June 6, 1918—the same day and month as the invasion of Europe on D day, 26 years later.

His Honor Was Right

Imprisoned by the Austrians as a spy, Lenin was saved by the mayor of Vienna, who thought he might prove useful as an agitator in Russia. How right he was! Sent across Germany in a sealed train, Lenin went home to lead Russia out of the war, giving Germany a handsome—though short-lived—peace settlement and most of the Ukraine!

The Best-laid Plans...

A courier's mistake brought about one of Germany's greatest victories of World War II.

On January 10, 1940, Maj. Helmut Reinberger decided to disobey orders and fly from Loddenheide airfield near Munster, Westphalia, to Cologne with a batch of top-secret maps showing Germany's planned attack on France. But the plane got lost and crash-landed in Belgium. The maps were seized and the French were alerted. Four months later the Germans came up with and executed a better plan—sweeping through the Ardennes and mopping up the French and British in just over four weeks,

with only about 150,000 casualties.

Ironically, the original scheme dated from World War I and would have cost the Germans many more lives!

Without Honor in His Own Country

German alchemist Fritz Haber invented a method of making explosives from nitrates, which helped Germany stay in World War I for two more years. He also headed the German program for the development of poison gas. His reward? He was fired and exiled by Hitler for being a Jew.

World War II: History Repeats Itself

"Those who cannot remember the past are condemned to repeat it," said Santayana. The world apparently didn't learn its lesson from World War I—and just 21 years later was plunged into another devastating global conflict.

But one man saw the menace of Hitler's Germany and issued warning after warning before war broke out in 1939. His name was Winston Churchill, a man who had been in the political wilderness for years. But in 1940 and on the brink of defeat, the British turned to him—and he chanced to be the inspirational leader they needed to guide them to victory.

Oops!

The Soviet legation in Helsinki was hit by the first bomb dropped by Russian aircraft when they attacked Finland on November 30, 1939.

The Unluckiest German Spy

Peter Karpin, a German spy in World War I, was captured by the French as soon as he entered the country in 1914. But the French sent faked reports in his name to Germany—and intercepted his wages and expense money until the agent escaped in 1917. With that money, someone on the French side bought a car and was driving in the occupied Ruhr when he knocked down and killed a man. It was Peter Karpin!

Dunkirk: A Lucky Escape

The British Expeditionary Force was trapped in France in 1940. The Germans were closing in and most of the ports were overrun. But luck, calm weather and a strange decision by Adolf Hitler all went in the Britons' favor and they were able to save 80 percent of their troops and 100,000 French soldiers.

The last suitable port left was Dunkirk, but Hitler ordered his troops to halt and not to enter it. Shrouds of fog blanketed the beach and the Channel was calm enough to enable thousands of small craft to cross over from Britain and bring the boys back home. It was the turning point of the war.

Voice from the Grave

The tomb of the emperor Tamerlane in Samarkand was opened by Soviet scientists at 5 A.M. on June 22, 1941—despite an ominous inscription which warned: "If I should be brought back to earth, the greatest of all wars will engulf the land."

At the same moment 2,500 miles to the west, the Germans were invading Russia in the Barbarossa campaign—the greatest armed conflict the world had seen to that date!

The Unsurprise Attack

Pearl Harbor, now famous as the "surprise attack" that brought the United States into World War II, was in fact not that much of a surprise! Several people in the American command knew it was going to happen; all that was missing was the exact place and time. And even this information came to light—but the message was delivered too late, thanks to chance slip-ups in the chain of command.

* There was plenty of evidence that the Japanese would soon enter the war. The U.S. had broken the Japanese code system and knew a two-part message declaring war would be delivered to Japanese envoys in Washington. The first part had arrived, and they were waiting for the second.

* Pearl Harbor wasn't placed on a war footing because it was halfway across the Pacific and seemed an unlikely target for an attack.

* Early on the morning of December 7, the day of the attack, a submarine periscope was spotted heading towards the harbor. Battle stations were called for an hour and then dropped when there were no more sightings. There were, in fact, more than two dozen Japanese submarines in the area.

* The submarine net around the harbor was opened to let in two U.S. minesweepers—and left open!

* The destroyer *Ward* sank a submarine and thought it had destroyed another, but the report was dismissed by people onshore.

* A radar station picked up Japanese planes flying missions over the north of the island, but the message was not decoded because everyone was having breakfast.

* At 3 P.M. that day, after the attack, a cable from Washington was finally delivered to the commanding officer, warning of a Japanese announcement and putting the base on the alert. It had been delayed in decoding and further delayed because it was a Sunday. It had been received at Pearl Harbor at 7:33 A.M.—almost half an hour before the first Japanese planes struck!

Bullet That Put Out a Fire

During the attack on Pearl Harbor, a Hawaiian Airlines DC-3 plane was hit by a Japanese explosive shell and burst into flames. A minute later, a machine-gun bullet hit the valve of the fire extinguisher—and put out the fire!

Pacific War Prophesied

In *Banzai*, written by German novelist Ferdinand H. Graetoff in 1908, a Pacific war is forecast. Unprepared U.S. troops under a General McArthur are at first beaten by the Japanese but later rally and win the war. Just 34 years later, it all came true!

Winds of Change

The famous kamikaze pilots of World War II, who committed suicide by crashing their planes into enemy ships, were named after the *kamikaze* or "divine wind" which saved the Japanese from the Mongols when a hurricane wrecked the invading Mongol fleet.

Blown—to Safety

Roy Dikkers was on board a tanker in December 1943 when three torpedos struck. The first jammed the door to his quarters, the second broke it open. He reached the deck—to find himself surrounded by burning oil. There seemed no hope until the third torpedo blew him into the air and clear of the burning waters—right next to a life raft. After drifting for three days, he was rescued by a Norwegian ship!

IN 1943 M/SGT. JOHN HASSEBROCK of Buffalo Center, Iowa RECEIVED A 3-DAY PASS TO MARRY A WAC CORPORAL WHEN HE WENT OVERSEAS. THEY LOST TRACK OF EACH OTHER UNTIL ONE NIGHT IN FRANCE HE MADE A CONVOY TO THE FRONT LINES AND WENT TO A FARM HOUSE TO SPEND THE NIGHT *THERE HE MET HIS WIFE!—ON THE EXACT DAY AND HOUR OF THEIR FIRST WEDDING ANNIVERSARY!*

Chance Saves the Fuehrer

Chance Saves the Fuehrer

In 1944 several of Hitler's officers hatched a plan to kill the fuehrer. Carrying a bomb in his briefcase, Count von Stauffenberg carefully placed it under the table near Hitler just before a conference at the "Wolf's Lair" in East Prussia on July 20, 1944. But another officer reached down and moved the briefcase so that a heavy slab of oak came between Hitler and the bomb. It exploded—killing four and injuring several others. But Hitler, who was leaning over the massive oak table at the time, was merely singed and deafened —the explosion cracked his eardrums.

The plot was quickly foiled, the ringleaders hunted down—and the war went on for another year.

A Close Shave

The V-2 rocket was perfected by the Germans in the final months of World War II. But they had a winged bomb much earlier. In 1943, a British ship in the Bay of Biscay had a narrow escape when one of these "guided missiles" was turned aside—by the noise of an electric razor. The British immediately equipped all their ships with such shavers and the Germans shelved the project.

Soviets Rocket Ahead

Soviet technology got an unexpected boost in the last days of World War II—when several V-2 rocket prototypes fell into Russian hands by mistake!

The Americans had given orders for 100 models to be taken from the Peenemunde rocket station, but the officer in charge wasn't told what to do with the rest. So, unbelievably, he left them for the Russians, who were officially occupying the area!

Old Soldiers...

Gen. Douglas MacArthur was fired by Harry Truman in a dispute over the handling of the Korean War in 1951. His father was fired by President Taft for insubordination. And one of his grandfathers defied orders in the Battle of Missionary Ridge in the Civil War!

The Lincoln-Kennedy Connection

Fate was playing some strange and cruel tricks the day President John F. Kennedy was assassinated in Dallas. Consider these eerie parallels with the murder of President Abraham Lincoln:

* Both Kennedy and Lincoln were deeply concerned with civil rights and were assassinated in office during years of great turmoil over civil rights—Lincoln over slavery, Kennedy over segregation.

* Both were shot on a Friday. Both wives were present.

* Both were shot from behind and in the head.

* Both were succeeded by men named Johnson, both of whom were Democrats from the South, one born in 1808 and the other in 1908.

* Lincoln's assassin, John Wilkes Booth, was born in 1839. Lee Harvey Oswald was born in 1939.

* Lincoln had a secretary named Kennedy who told him not to go to the theater that night. Kennedy's secretary, named Lincoln, warned him not to go to Dallas.

* Booth shot Lincoln in a theater and ran into a warehouse; Oswald shot Kennedy from a warehouse and ran into a theater.

* Both presidents' names contain seven letters; their successors' names have 13; their assassins' have 15.

PART EIGHT:

Through a Glass Darkly

Abracadabra and the Hand of Destiny

Man has always sought to control chance.

His magic, superstitions, religions, laws and even governments have been established to assure him a place in the universe. Whether through astrology or palmistry, biorhythms or fortune-telling, charms or talismans, dreams or prophecy, man has always tried to persuade himself that nothing belongs to the force of chance.

In this chapter, we examine man's efforts—some futile, some remarkably successful, some ridiculous and some brilliant. We look at the curious coincidence of time-twins and find the story of a king and a commoner, both inextricably wed by the planets.

There are fatal astrological predictions and favorable signs, the curse of the full moon plus lucky days and numbers.

We learn of superstitions and sailors' curses, of death ships and voyages that never ended. We read of dreams that changed history, of books that foretold events years before they happened.

There are stories of kings and presidents, queens and empresses, princes and princesses. We find Abraham Lincoln was connected all of his life with the number 7.

We meet witches, alchemists, magicians, and the psychics who solve crimes. We look at religious miracles, your future in a teacup, the science of biorhythms and how they affected everyone from John Kennedy to Evel Knievel.

There are superstitions, strange dreams, the black arts and the curse of the Bermuda Triangle. It is impossible for a human being to turn instantly to dust, but we learn of it happening.

We swing open the doors and show the wisdom—or the folly—of controlling chance. Can it be done? You decide.

A Starry Chance

Chances are if you read a daily newspaper, you will see a column listing your horoscope by a sun sign. Is astrology chance or is it a scientific law of the stars?

We turn now to this fascinating world where some swear by their horoscope, others produce documented data to support sun signs and still others plot their lives around a horoscope and good or bad days.

Some stories defy the laws of chance. Others may seem too farfetched to be true. Judge for yourself as we visit the world that predicts what tomorrow will hold.

Does It Work?

Is astrology a science or is it just based on chance predictions?

∗ Queen Victoria and composer Richard Wagner were both born with the sun rising in Gemini, a trait that produces compulsive writing. Wagner wrote operas and Queen Victoria was a compulsive letter writer, filling her desks with constant notes on everything.

∗ Writer Jules Verne has a birth chart showing Mercury at midheaven in Aquarius. That has always meant a strong interest in science plus a creative bent, producing a prolific science fiction author.

∗ Isadora Duncan has Venus in the 12th house in Aries, a combination that meant she would be sensual. She demonstrated that in later life with her dancing and scandalous love affairs.

∗ Marilyn Monroe had the planet Neptune rising in Leo in the first house—a chart that astrologers say shows high ideals, sentimentality, ambition and pushiness. Many say this is why she was constantly at odds with herself over what she wanted in life.

Pick a Number

Twelve is the most important number in astrology. A horoscope is charted into 12 houses, the zodiac has 12 signs. To prepare a horoscope, an astrologer needs the correct time of birth to within 30 minutes, the latitude and longitude, the day, month and year.

Once he has calculated these positions, he can then attempt to eliminate chance in your life. He might tell you your future, your strengths and weaknesses, dangers to your health and even the type of person you'll marry.

The zodiac signs and some of their characteristics in people:

Aries (March 21–April 20): Radiant, full of energy and headstrong.

Taurus (April 21–May 21): Stubborn, creative and wealthy.

Gemini (May 22–June 21): Changeable, quick-witted and versatile.

Cancer (June 22–July 23): Sensitive, moody and giving.

Leo (July 24–Aug. 23): Generous, overbearing and enterprising.

Virgo (Aug. 24–Sept. 23): Discriminating, fussy and efficient.

Libra (Sept. 24–Oct. 23): Refined, indecisive and aloof.

Scorpio (Oct. 24–Nov. 22): Imaginative, jealous and willful.

Sagittarius (Nov. 23–Dec. 21): Sociable, adventurous and juvenile.

Capricorn (Dec. 22–Jan. 20): Reliable, pessimistic and authoritative.

Aquarius (Jan. 21–Feb. 19): Idealistic, unpredictable and free.

Pisces (Feb. 20–March 20): Compassionate, melancholic and sometimes mystical.

The Signs Are

Astonishing coincidence or reality? The arguments still rage, but history shows the importance of astrology, and today 50 million people in North America read their horoscope each morning.

It's a 2,000-year-old art, but astrologers point to the following to illustrate more than chance historical happenings:

Dr. John Dee

* The discovery of the planet Uranus by scientists in 1781 heralded the French Revolution eight years later.

* Neptune was found in 1846 and shortly after there was a huge increase in spiritualism and psychology.

* Pluto was discovered in 1930 and many feel this heralded the Depression, World War II and the atom bomb.

* Both the Germans and the Allies used astrologers during World War II. Hitler and his Third Reich published an astrology magazine predicting great victories. The Allies countered this by dropping their own magazines across Germany predicting disastrous fates. D day was planned, not just because of the weather, but because the planetary positions were favorable.

* Elizabeth I ruled England with the help of her personal court astrologer, John Dee, who worked out the days and times to launch invasions, repel the Spanish Armada and send out explorers.

* St. Thomas Aquinas included in his religious teachings the edict that the planets cause "all that takes place in the sublunar world."

* The great pharaoh Rameses II consulted his personal astrologer before starting a temple.

* Hippocrates thought that any doctor who didn't know astrology had no right to practice.

Hocus Pocus

Synopsis of the diviner's arts, including chiromancy, prophecy, minemonics and, centre right, geomancy and the geomancer's shield.

The Job Market

Some astrologers have honed their art to a degree where they can eliminate chance in job hunting by predicting the occupation best suited for your sign.

For example, this list shows jobs, success and failures by signs:

Occupation	Most Likely	Least Likely
Journalists	Scorpio	Capricorn
Librarians	Libra, Scorpio	Capricorn, Aquarius
Authors	Virgo	Gemini
Scholars	Cancer, Libra	Scorpio, Capricorn
Athletes	Leo	Aries
Pro ball players	Libra	Gemini
Businessmen	Cancer	Sagittarius
Singers	Aquarius	Virgo
Composers	Capricorn	Virgo
Musicians	Aquarius, Pisces	Cancer, Taurus
Actors	Taurus	Cancer
Celebrities	Taurus	Scorpio
Painters	Cancer	Pisces
Architects	Capricorn	Pisces
Engineers	Cancer	Libra
Scientists	Capricorn	Leo
Lawyers	Gemini	Pisces
Clergy	Taurus	Capricorn
Teachers	Leo, Virgo	Capricorn
Doctors	Leo, Libra	Aries, Cancer, Sagittarius
Bankers	Virgo, Aquarius	Pisces, Taurus
Politicians	Cancer, Libra	Aquarius, Aries

Stars and Their Stars

Since astrology measures only tendencies, you can interpret signs in your own way. But documented birth signs show:

∗ The day Charles II was born in 1630, a comet was seen over Britain.

∗ Virgin Queen Elizabeth was born on the eve of the feast of the Virgin Mary.

∗ John Dryden read his son's birth chart and predicted disasters at the ages of 8, 23 and 33. At 8 the boy was involved in an accident with a stag, at 23 he fell from a tower and at 33 he drowned.

∗ King George III and an ironmonger named Samuel Hemming were both born on June 4, 1738, at the same moment in the same town, making them astrological time-twins. Both married on September 8, 1761, both had nine sons and six daughters, both became ill and had accidents at the same time and both died on January 29, 1820.

∗ Albert Einstein and Otto Hahn, both born on March 14, 1879, both won the Nobel Prize in physics.

∗ Two astrologers predicted that Prince Charles and Lady Diana would have a baby within a year of marriage. They also predict two more children within the next seven years.

Psst, Wanna Buy a Sign?

Chances are good you can make money if you turn to astrology.

A firm on Long Island sold more than 250,000 horoscopes in 1969, and a book about the Washington sybil Jeane Dixon sold 260,000 hardback copies and nearly three million paperbacks in 1970.

There's also documented proof that many businessmen are obsessed with peering into the future. Some even retain private astrologers to get dates for favorable takeover bids, mergers and movements in the stock market.

Modern Conclusions

Even hard-nosed scientists say there is more to astrology than chance, especially during certain periods of the moon.

The number of suicides and crimes of passion increases dramatically during a full moon; emergency wards double staff in high-crime areas when the moon is full; and the behavior of rats in confined spaces becomes much more violent when a full moon is shining.

Russian scientists, meanwhile, have found that there are four times as many accidents the day after a solar flare or sunspot than at any other time.

And engineer John H. Nelson produced documented evidence that the earth's magnetic field is disturbed when major planets change conjunction.

That Explains It!

Time now for a peak at the horoscopes of some famous people. All the comments were written about their sun sign, so we judge for ourselves whether it was chance or science:

* Richard Nixon, J. Edgar Hoover, Mao Tse-tung (Capricorn): "Strong, highly organized and preoccupied with personal prestige."

* Franklin Roosevelt, Adlai Stevenson, Abraham Lincoln (Aquarius): "Tolerant, affectionate, loyal, live in a world of dreams."

* Patty Hearst, Ted Kennedy, Elizabeth Taylor, Jackie Gleason (Pisces): "The least stable and most interesting of the signs—constantly changing in response to inner moods and external pressures."

* Nikita Khrushchev, Marlon Brando, Bette Davis (Aries): "Petulant, impatient, self-centered, they want their own way."

* Harry S. Truman, Sigmund Freud, Oliver Cromwell (Taurus): "Stubborn, argumentative, unchanging, determined, self-reliant."

* J. F. Kennedy, John Dillinger, Brigham Young, Errol Flynn (Gemini): "Youthful and energetic, capable of taking either side of the argument, physical, even failure can be spectacular."

* Calvin Coolidge, Gerald Ford, Andrew Wyeth (Cancer): "Stable, secure, dedicated to family and friends, highly protective of status."

* Jacqueline Kennedy, George Bernard Shaw, Napoleon Bonaparte (Leo): "Have firm and intense ideas on how the world should be run."

* Lyndon Johnson, H. L. Mencken, William Howard Taft (Virgo): "Savers of string and tireless solvers of every problem they encounter."

* Dwight Eisenhower, Mahatma Gandhi, Charlie Brown (Libra): "Always in the middle, logical, harmonious, can take one side or the other with equal ease."

* Richard Burton, Margaret Mead, Robert Kennedy (Scorpio): "Intensely personal and physical, constantly searching for answers, careful to hide their own feelings."

* Frank Sinatra, James Thurber, Mark Twain (Sagittarius): "Frank to the point of being rude; honest and easygoing."

Luck

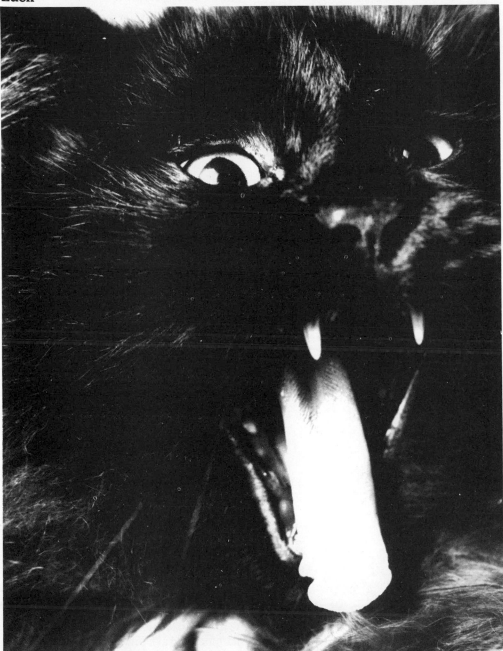

Scared of black cats? Worry on Friday the 13th? Avoid walking under ladders? Well you're not alone.

Surveys show that 50 percent of the people in the world believe in some form of luck and 24 percent believe in lucky numbers. Come with us now as we explore some of the more popular theories on luck, superstition and blind chance.

Lucky Days Are Here Again

People still believe in lucky and unlucky days, a belief that goes back to Mesopotamia.

It's considered unlucky to begin anything on a Friday. You should not get married, move house, start a new job, begin a long journey or wean a child. Nor should you cut your nails or turn a mattress.

Professors today say the unlucky Friday can be traced back to the crucifixion of Christ, which happened on a Friday. Thus the day is considered a time for endings, not beginnings.

Sunday is considered a lucky day on which to be born but unlucky for most everything else. It is considered the worst possible luck to bury someone on a Sunday, because it is supposed to be set aside for churchgoing and rest.

Saturday is considered the correct day for digging up or disposing of vampires in Bavaria, but was considered unlucky for weddings in the 15th century. This is largely

ALFRED GWYNNE VANDERBILT

THE AMERICAN MULTIMILLIONAIRE WAS SO SUPERSTITIOUS THAT, TO WARD OFF EVIL SPIRITS, HE ALWAYS SLEPT WITH THE LEGS OF HIS BED RESTING IN *DISHES OF SALT* ··· HE DIED IN 1915 ON THE TORPEDOED LUSITANIA— *DROWNING IN SALT WATER*

ignored today because Saturday is the most convenient day.

One marriage tradition still lingers in Europe, however. May is considered an unlucky month for weddings because of the Roman custom of making offerings to the dead during that month.

The Old Rabbit's Foot

Charms, amulets and talismans are intended to attract good luck and ward off bad. Many hang a horseshoe over a door, carry a rabbit's foot or wear a charm bracelet. In Europe, a nail is a luck-bringer and so is a lump of coal which carries the warmth of fire.

There are others that defy chance and explanation:

* The Egyptians were addicted to amulets and two survive today—the scarab and the ankh. The scarab is a symbol of the sun as a source of life and the looped-cross ankh symbolizes the force of life.

* The Polynesians have a tiki, a small carving of a human figure associated with birth. It also has become a popular North American charm.

* Red coral is still attached to baby rattles in some countries to ward off the evil eye, while red jewels and semiprecious stones are widely used as amulets.

* Astrological talismans are usually made of the metal associated with a planet and manufactured at a time when the planet is dominant. A silver talisman, the metal of the moon, is supposed to bring good health and respect when carried or worn.

A lead talisman, the metal of the ominous Saturn, is considered a bad-luck charm when buried near a house.

Witches, Magicians and Alchemy

During the Middle Ages and the early Renaissance in Britain and Europe, witch-hunting was a horrible pastime. The targets of hunters of witches, magicians, seers and sorcerers were usually the old and the helpless, the friendless or the too beautiful.

Chance and coincidence usually formed the largest part of the evidence at any witch trial. The frenzy even spead from Europe to the infamous Salem witch-hunts in North America.

* In 17th-century Austria it was possible for a man to accuse a woman of witchcraft if a piece of her clothing was found in his room. The witches were usually questioned under torture and then burned, whether they confessed or not.

* In Scotland in 1696, one spiteful little girl named Christine Shaw caused the execution of 21 persons as witches. She decided it was a game to act as if she were possessed and then to accuse the two prettiest local girls of trying to kill her. Local witch-hunters soon got in the act and she denounced everyone they suggested.

* Laws in France during the 16th century gave the witches immunity if they accused someone else; forced children to testify against their parents; protected the names of the accusers; and allowed torture on suspicion. The laws came from a pious Carmelite monk who also decreed an accused person could never be acquitted.

* Two old women were arrested, tortured and burned as witches in Bury St. Edmunds, England. One was a baby-sitter who tried to soothe a child and the other cursed at a cart driver who broke her window.

* Witches in Offenburg, now the modern city of Baden, were burned for their money and possessions which were confiscated by the priests and town council. The practice became so widespread that from 1628 to 1630, 79 persons out of a population of 2,000 were burned.

Chances of Being a Witch

You could have been judged a witch in the Middle Ages if a pet disobeyed you or you floated in water. A mole or wart on the body was also considered a curse and neighborhood children could accuse you if they didn't like you.

Jews, however, could not be accused of witchcraft because they were not Christian.

The Great Magician

One of history's most famous magicians still has prophecies coming true 400 years after his death.

Michel de Notre Dame was a skilled prophet and healer, but his study of medicine and astrology earned him the name Nostradamus and the ears of kings and queens.

He forecast the 1930s' rise of a dictator in Germany who would cause a world war and fight without conscience. He also predicted the French Revolution, the power of Napoleon and the deaths of kings and queens.

He forecast his own death, including the date, place and time.

Nostradamus told Catherine, queen of France, that her son would die before his 18th birthday and then predicted that her youngest son, Henry of Navarre, would succeed her. Both prophecies came true.

One of his last predictions involves the year 1999. He said a "king of terror" will descend from the clouds in the month of July and rule the world. The prophecy is felt by many to mean an invasion from outer space.

The Great Lover

The world's finest lover made many of his conquests because he was skilled in magic and the occult.

Casanova, the son of an actor, quickly realized magic was theatrical, so when he saved the life of a senator by removing a filthy bandage, he proclaimed he had done it with magic.

That established his reputation and he continued his conquests among females, telling one her pregnancy would only end with the insertion of fresh male sperm—his own....

The Marching Ants

A psychiatrist obsessed with the connecting principle in luck proved his case with thousands of ants.

C. G. Jung devised the term synchronicity to describe the phenomenon of two people who had a series of chance encounters on a street. The connector is more than luck, he wrote about this particular example, because even though the streets were different, they always met before a house with a green door.

To prove his theory, he took three matchboxes and filled the first with 1,000 ants, the second with 10,000 and the third with 50. He bored a hole in each box and added one white ant in each with the black ones.

In every case, the first ant to come out of the box was always white. A feat that defies chance?

The FATHER WHO SENSED
HIS SON'S DEATH
WILLIAM W. OLIVER (1778-1869)
COLLECTOR OF CUSTOMS IN
SALEM, MASS., FOR 46 YEARS
REFUSED TO MEET THE SAILING SHIP,
"GEORGE," ON WHICH HIS 18-YEAR-OLD
SON, WILLIAM G. OLIVER, HAD
EMBARKED FROM CALCUTTA, SAYING:
" There is no sense in
meeting her. My son is dead!"
THERE WAS NO POSSIBLE WAY IN
WHICH THE NEWS COULD HAVE
REACHED OLIVER, YET HIS SON
HAD DIED ABOARD THE SHIP 2
MONTHS BEFORE IT REACHED SALEM

The Winning 10

Everyone likes to take a chance on a lottery, but how many know they're going to win— even if the odds are 63 million to 1 against them?

Eric Leek won one of the largest lotteries in history in 1976 when he cashed the winning ticket on the Bicentennial lottery in New Jersey.

And he says he knew he was going to win, all because of the number 10. Leek was born on the 10th hour of the 10th day of the 10th month. His ticket had a 10 on it. He was number 10 on the list picked and the final draw was the 1st month, 27th day, three digits that add up to 10.

Leek was right and picked up $92,000 a year for life, plus a guarantee of $1.8 million if he dies early.

Leuben

Famous German Lunatic
BET THAT HE COULD
: TURN UP A PACK OF CARDS
IN A CERTAIN ORDER

HE TURNED THE CARDS
10 HOURS A DAY
FOR 20 YEARS
EXACTLY 4,246,028
TIMES
BEFORE HE SUCCEEDED

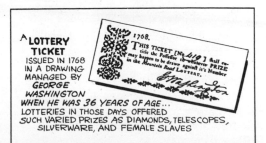

A LOTTERY
TICKET
ISSUED IN 1768
IN A DRAWING
MANAGED BY
GEORGE
WASHINGTON
WHEN HE WAS 36 YEARS OF AGE...
LOTTERIES IN THOSE DAYS OFFERED
SUCH VARIED PRIZES AS DIAMONDS, TELESCOPES,
SILVERWARE, AND FEMALE SLAVES

1768.
THIS TICKET [No. 419] shall entitle the Possessor to whatever PRIZE may happen to be drawn against it's Number in the Mountain Road Lottery.
G Washington

Get Lucky

Conrad Hilton.

Want to improve your luck? Scientists and psychiatrists say studies of the lucky show:
* They form a lot of friendships. Kirk Douglas, for example, was friendly all his life and got his break in the film *Champion* because of help from actress Lauren Bacall who remembered him as a likable young actor.
* Back your hunches. Conrad Hilton put in a $165,000 bid once for a Chicago hotel, but then had a bad dream about not getting it. He changed his bid the next morning to $180,000 and got his new hotel—by only $200.
* Be bold. J. Paul Getty, the oil billionaire, took a chance while an art student and became an oil wildcatter. After three losses, he hit it lucky at age 23.
* Limit your losses. Gerald Loeb, one of the luckiest stock market speculators, said his secret was knowing when to sell and "having the guts to do it."
* Prepare for problems. A study of auto accidents shows the most "hard-luck" types turned out to be people who were overly optimistic.

Luck on the High Seas

No one is more superstitious than a sailor. The albatross and the red sky at morning are both signs that seamen obey. But bad luck still curses them:

* The steamer *Père Marquette* was wrecked in 1888 on Lake Michigan, sank in 1890, was refloated and went aground in 1893. In 1902 it went aground again and in 1916, it collided with a bridge. It was caught in an ice field in 1920 and finally junked as a cursed ship in 1921.
* The C.S. *Holmes* was being swept towards the rocky coast of Vancouver Island during a gale in 1909, but was miraculously saved by a change in the wind. Exactly 41 years later, another gale wrecked the ship in exactly the same position.
* The Mississippi river packet *Jo Daviess* is considered one of the worst jinx ships in history. It sank after three trips and its engines were installed in the steamboat *Reindeer*, which sank after four trips. They then went to the *Reindeer II*, which also sank after four trips. The motors then went to the *Colonel Clay*, which sank after two trips, and then to the S.S. *Monroe*, which was destroyed by fire. The engines were then taken ashore and put in a grist mill in Elizabethtown, Pennsylvania. A year later it burned to the ground.
* The *Erinna*, a Nova Scotian bark, made 16 voyages in seven years and at least one member of the crew died on each trip. On February 3, 1885, another crew member drowned and the *Erinna* was abandoned at sea when its captain went mad.

THE **CURSING STONES OF TORY ISLAND,** Eire,

LEGEND HAS IT THAT TURNING THE STONES COUNTER-CLOCKWISE CAN PROVOKE DISASTER -- IN 1884 "THE WASP," A GUNBOAT, WAS SENT TO PUNISH A TAX REVOLT - - ISLANDERS TURNED THE STONES *AND THE SHIP WAS WRECKED!!*

Numerology: Chance or Science?

The chance pairing of letters and numbers dates back to the Greeks and Romans, who were obsessed with divining their fates.

Today, numerology is practiced everywhere in prophecy, prediction and astrology. The simplest method is to write down the numbers 1 to 9 and then under them write the letters of the alphabet.

1	2	3	4	5	6	7	8	9
A	B	C	D	E	F	G	H	I
J	K	L	M	N	O	P	Q	R
S	T	U	V	W	X	Y	Z	

Write your own name, or nickname, and add up the assigned numbers. If the total is more than 9, add the digits together to get a single number (54 is 9 for example) and you have a key to your overall personality.

Ones are born leaders, single-minded, ambitious and determined. But they are also considered pushy, egotistical and have little regard for anyone else in a drive to power.

Twos are followers, weak and indecisive. On the positive side, they are tranquil, good listeners and very agreeable.

Threes are lucky, charming and attractive. On the other side, they are careless and take their good luck for granted.

Fours are usually failures. They plod along, are bad lovers and generally don't stand out. But they are also considered hard-working and industrious.

Fives are attractive, charming and sexy. Negatively, they can be vain and inconsiderate.

Sixes are homebodies with good qualities of motherlove and family. They can also be uninspired and dull.

Sevens are mystics. They have an imagination and interest in the occult. But they tend to be secretive and introverted.

Eights are great successes in business and finance, but are sometimes too ruthless.

Nines are spiritual and political leaders, single-minded and inspired. They have few negative features.

Does It Work?

Judge for yourself. Many believe religiously in numerology, but others say it is only coincidence. Yet:

* Napoleon Buonaparte was quite a success as a 1 until he decided to drop the *u* in his last name to make it more French. After he did it, he never won another battle and ended up an exile on St. Helena where he died a broken man. By dropping the *u*, he had made himself into a 4.

* Norma Jean Baker adds up to a 6, a homebody, a good wife and mother. Hollywood changed her name to Marilyn Monroe a 2, the right number for a female sex symbol.

* David O. Selznick changed from the struggling son of a producer to a movie mogul after he added the *o* to his name. But he didn't believe in numerology, he just went through the alphabet until he found a letter he liked.

THE FATAL NUMBER

The FOLLOWING LIST OF KINGS HAVE EITHER BEEN MURDERED OR DEPOSED

NICHOLAS	II	— of RUSSIA
WILLIAM	II	— of GERMANY
JAMES	II	— of ENGLAND
FRANCESCO	II	— of SICILY
CHARLES	II	— of FRANCE
CHARLES	II	— of ENGLAND
CHARLES	II	— of ANJOU
FREDERICK	II	— of GERMANY
WILLIAM	II	— of ENGLAND
ALEXANDER	II	— of RUSSIA
MANUEL	II	— of PORTUGAL
JEAN	II	— of FRANCE
RICHARD	II	— of ENGLAND
PETER	II	— of RUSSIA
HAROLD	II	— of ENGLAND
FRANZ	II	— of GERMANY
EDWARD	II	— ENGLAND
ETHELRED	II	— of ENGLAND

Fate by Number

Numerologists say you can also work out your own fate by adding up your birth date. Use the month, day and year.

Certain numbers are associated with being lucky, some less so. 1, 3, 5, 7, 9 are all considered lucky; 4 is the least lucky.

And across the world, some numbers are lucky—or unlucky—in certain countries:

* 13 is considered unlucky in North America, and being the 13th at a table is considered a curse. At the last supper, the 13th was either Christ or Judas.

* 2 was considered unlucky by the Hebrews who would never do anything twice. In their number system, 1 stood for God and 2 the Devil. There are also two forks in the tongue of the serpent, duplicitous people are referred to as two-faced and the Roman god Janus, the bitter enemy of the Hebrews, had two opposite faces.

* 4 is considered to be the number of solidity and, in some countries, the basic foundation of the world. There are four seasons, four weeks in a month, four elements (earth, air, fire, water) and four states of being (hot, cold, wet, dry). There are also four gospels, four phases of the moon, four wheels on a car, four walls in most rooms and four legs on a table.

* 7 is the most magical number of all and is universally regarded as an important number in religion and magic. In Buddhism, there are seven points on the body through which the spirit may depart.

In Europe, the seventh son of a seventh son is said to have magical powers. There are seven days in a week, seven days of creation in the Bible, and each phase of the moon lasts seven days.

According to some, the body renews itself every seven years. Asthma runs in seven-year cycles, a woman's monthly cycle is four groups of seven days, there are seven openings in the body, seven notes in the musical scale, seven deadly sins, seven wonders of the world, seven requests in the Lord's Prayer, seven colors in the spectrum.

* 3 is also considered lucky because good things come in threes. In Scotland, however, they say death comes in threes. All time is in three parts—past, present and future. In the fairy tale there are three wishes and three metals are used for coins. There are three fates, three golden apples, three measures (length, breadth and width) and most folk heroines had three suitors. There are also three parts to the Holy Trinity and the riddle of the Sphinx. People still say "I'll give you three guesses," and the Greeks thought it the perfect number because it made a triangle and had a beginning, middle and end.

The Devil's Messenger

The number 666 has long been used to describe Satan's messenger on earth. It is associated with the great beast of Revelation. Chance or not, consider:

* It was assigned at one point to Napoleon because there are 18 letters in his whole name or three groups of six.

* Jehovah's Witnesses believe it stands for the United Nations.

* Many used it to describe the British House of Commons with 658 members, four clerks, a sergeant, a deputy, a doorkeeper and a librarian—666!

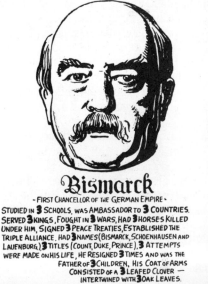

Bismarck
- FIRST CHANCELLOR OF THE GERMAN EMPIRE -
STUDIED IN 3 SCHOOLS, WAS AMBASSADOR TO 3 COUNTRIES, SERVED 3 KINGS, FOUGHT IN 3 WARS, HAD 3 HORSES KILLED UNDER HIM, SIGNED 3 PEACE TREATIES, ESTABLISHED THE TRIPLE ALLIANCE, HAD 3 NAMES (BISMARCK, SCHOENHAUSEN AND LAUENBURG) 3 TITLES (COUNT, DUKE, PRINCE), 3 ATTEMPTS WERE MADE ON HIS LIFE, HE RESIGNED 3 TIMES AND WAS THE FATHER OF 3 CHILDREN, HIS COAT OF ARMS CONSISTED OF A 3 LEAFED CLOVER — INTERTWINED WITH 3 OAK LEAVES.

The Lincoln Story

Few tales of numbers are stranger than the story of Abraham Lincoln.

Each of his names contains seven letters, he lived seven years in Kentucky and seven years in Salem, he was a private (seven letters) and a captain (seven letters), he was elected seven times, sworn into Congress December 7, 1847, held seven offices in succession, had seven debates with Stephen Douglas (twice seven letters), fought slavery (seven letters), his ancestors came from Hingham (seven letters), England (seven letters), county of Norwich (seven letters), moved at the age of 7 to Spencer (seven letters) in Indiana (seven letters), was seven years in state legislature, appointed seven cabinet ministers, watched seven states secede and died a few minutes after seven on the seventh day.

Good Things Come in Threes

Thomas Jefferson was born April 13, 1743, the third child and the third Thomas in his family. He wrote the Declaration of Independence at the age of 33, was the third commissioner, the third member of Washington's cabinet, the third president of the American Philosophical Society and the third ambassador to France for three years.

He lost the presidency in 1796 by three votes but was elected third president of the United States in 1800. He died at the age of 83.

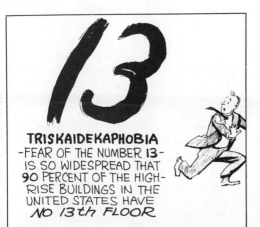

TRISKAIDEKAPHOBIA
-FEAR OF THE NUMBER 13-
IS SO WIDESPREAD THAT
90 PERCENT OF THE HIGH-
RISE BUILDINGS IN THE
UNITED STATES HAVE
NO 13th FLOOR

They've Got Your Numbers

Time now to turn to some stories about numbers, all stranger than mere coincidence:

* J. M. Schwoob of Cody, Wyoming, was issued license plate number 1 for 17 consecutive years.

* Lola Susan Sutterby of Angola, New York, was born on the seventh day of the seventh month in the Seventh-day Adventist Hospital. She was born at 7 P.M., weighed seven pounds and was the seventh child in her family.

* Seven stockbrokers in Toronto, all working on the seventh floor of a 777-address building, went to the racetrack in 1977 on the seventh day of the seventh month and bet $777 on the seventh horse in the seven-furlong seventh race. The horse finished seventh.

* The numbers 19, 33, and 42 are unlucky in Japan because they sound respectively like the Japanese words for "repeated hard luck," "terrible trouble" and "death."

* Mrs. Walter Holman was born on the third day of the third month in 1933, the third daughter of a 33-year-old mother. She later married a third son and had three children.

* Crawford "Cherokee Bill" Goldsby, an Oklahoma outlaw who killed 13 persons, was captured for a $1,300 reward. Thirteen witnesses testified against him, the judge took 13 minutes to charge the jury, the trial lasted 13 hours, he was executed April 13, 1896, on a gallows that had 13 steps.

* Benjamin Franklin was born on January 17, 1706, was one of 17 children, started his career at 17 and died on April 17.

* Stan Gilbert of Glendale, California, on the eighth day of the eighth month became the grandfather of an eight-pound, eight-ounce boy. To celebrate, he went to Del Mar racetrack, bet $8 on the eighth horse in the eighth race at eight furlongs and won $80.80.

The Writing on the Wall

Longfellow, born in 1807, graduated at the age of 18 and became a professor in 1836 (the numbers 1, 8, 3, 6 total 18), married 18 years after graduating, and was a professor for 18 years. He was married to his second wife for 18 more years, published 18 volumes of poems and became ill on March 18, 1882.

Hands Up!

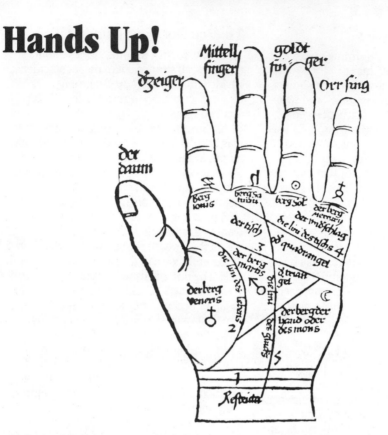

Your life, your fate, your future and your past could all be mapped out on the palm of your hand. If you believe in near-at-hand predictions, you can chart your life through the lines, mounds and shapes of hands and fingers.

Everyone has four lines corresponding to life, fortune, intellect and loves. According to their strength, length, shallowness or depth, a palm reader predicts the future.

There are other more specialized lines that show fame, health, intuition, travel, marriage and even children. The color of the palm, the temperature of the skin, the mounds at the base of each finger, the curve of the thumb, the distance between fingers, size and shape of nails—these are also considered.

But even if you don't believe in palmistry, consider this:

* A firm handshake usually means trust, a cold hand shows questionable motives. A wet or clammy handshake can mean you think less of the person you're meeting. A

meaty hand can make you insecure because you feel swallowed up.

* A man's hand with long slender fingers and smooth skin can mean one of two things: it may tell us that the man has never done a day's work in his life, or that he is a musician or surgeon.

* Doctors say florid palms mean high blood-pressure, cold hands poor circulation, pale yellow palms anemia.

* Does money slip through your fingers? In palmistry, it's a fact. Hold your hands up to the light with your fingers together. If you can see through them, it usually means you're a spendthrift.

* Laborers have squared fingers.

* Neurologists observe that children born with Down's syndrome have a distinct pattern of lines on their thumbs.

* In palmistry, a number of small lines on the palm indicates tension. Next time you have lunch with a hysterical friend, take a look.

The Fortunes of Magic

Cagliostro jumped from country to country throughout the 18th century, but his magic finally deserted him.

A skilled magician, forger and possessor of second sight, Cagliostro was forced out of Italy because he had a habit of predicting lottery numbers. In France, he became an enemy of Marie Antoinette and went to England after predicting the start of the French Revolution.

He eventually returned to Rome, but fell into the hands of the Inquisition and died in prison.

The Man Who Lived Forever

LE COMTE DE S.ᵗ GERMAIN
CÉLÈBRE ALCHIMISTE.

More than chance ruled the life of the mysterious Count St. Germain, the man Voltaire described as one who lives forever and knows everything.

The count spread the legend that he was hundreds of years old and had the secret to the fountain of youth. He gave his "rejuvenation elixir" to Madame de Pompadour and was a member of the Secret Brotherhood of Master Magicians.

He was seen in the U.S. Congress in the late 18th century at the age of 60 and again in the mid 19th century, without a sign of having aged.

The circumstances of his death and burial are still obscure.

Through a Glass Darkly

Rasputin is still thought by many to have been a black magician. He was the most powerful figure in St. Petersburg during the turbulent period prior to the Russian Revolution and despite tales of orgies, black masses and torture, he was a close adviser to the czar and czarina.

His life defied the laws of chance. He became a monk after a pilgrimage to the Holy Land on foot—2,000 miles each way. He offered his services to the czarina and three times healed her hemophilic son, once by praying while he was hundreds of miles away.

He also predicted the coming of the Russian Revolution, but added he would be murdered first. He warned that if he was killed by aristocrats, the revolution would go ahead. If he was killed by common people, the conflict would be avoided.

He was poisoned at dinner by a group of Russian noblemen, but they found it didn't kill him. They then shot him several times and when he still lived, they threw him into the river where he finally drowned.

The revolution started soon after and the czar and his family were killed.

Creatures of Chance

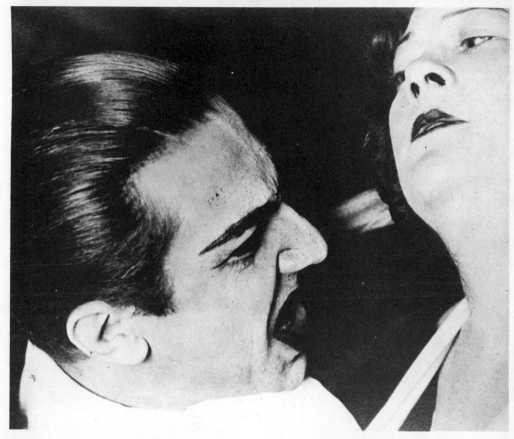

Where did tales of monsters, vampires, devil worshipers and witches first come from? Many feel they were old stories and legends rooted in superstition, but in fact most really existed and were even more horrible than a superstitious fairy tale:

* Countess Elizabeth Bathory of Hungary used to stay young by bathing in the blood of young innocent village girls. They were kept chained in her dungeon and were bled until they died. When she was finally arrested in 1610, about 50 bodies were found.

* Gilles de Rais, a French nobleman who was once the richest man in Europe, sexually molested and murdered 46 children, dismembering their bodies in satanic rites. He was finally arrested and executed—not for the murders, but for making a pact with the Devil.

* Dracula was based on Vlad Tsepesh, a Rumanian nobleman called Vlad Dracul or Vlad the Impaler. He got his nickname from his habit of impaling the heads of his enemies on pikes around his castle walls and once had a helmet nailed to a Turkish messenger's head.

* Matthew Hopkins was witchfinder general in 17th-century England and traveled around the country on his horrible missions. He determined guilt by finding moles or throwing witches into wells. If they floated, they had to be burned.

* King James VI of Scotland personally oversaw the trials of the North Berwick witches and helped burn more than 70 persons.

Alchemy

The search for gold has driven men and monarchs mad. All wanted to eliminate the element of chance in discovering this most precious metal, so many turned to alchemists in their obsession with turning base metals into gold.

Alchemy was really a sacred purification rite for old-style magicians, leading to the philosophers' stone, which could renew life, confer immortality and bring great wealth. It was a basic process of metal refining which was kept secret for many years after it was passed into Europe by the Arabs.

Each step in the alchemist's scale produced a stronger metal, but no one ever really turned it into gold. Yet that didn't stop:

* Charles II of England, who had an alchemical laboratory in his bedroom.

* James IV of Scotland, who not only believed in alchemy but thought he could fly and once broke both his legs when he jumped from his castle window.

* Maximilian II of Bavaria, who imported so many alchemists that he had to subdivide his house to hold their labs.

Psychic Stuff

Clairvoyants and psychics have been co-operating with police for years, but only in "unofficial" capacities. Law officials are still reluctant to admit psychic phenomena exist, but Dutch police openly consult psychics for assistance.

Even the most skeptical, however, must consider the following cases, which defy a chance prediction or guesswork:

* When Edward Hayward was shot in the Alberta woods by a greedy prospector in 1904, there were no witnesses—except for the victim's brother who was 6,000 miles away. George Hayward awoke in England from a dream in which he "saw" the murder so clearly that he went to Canada and told officials, who arrested the guilty man.

* Mentalist Professor Gladstone stopped in the middle of his act in a theater in 1930 and pointed to a man in the audience. "Your friend, Scotty McLaughlin, has been murdered," he said. He directed police to the murdered man's farm, described the death scene and even located the corpse.

* Patrolman Don Sabel of Grosse Pointe Woods, Michigan, got a radio call in his police car for a robbery. He started to answer it but then ordered the car around, told his partner to stop in front of a restaurant, jumped out and stopped a man. He found $500 from the robbery and the man quickly confessed. On another occasion, he had another "hunch" and brought a man in for questioning. The man became nervous and confessed to a robbery—even though the crime had not yet been reported to police.

* Psychic Edgar Cayce was consulted in a murder case and went into a trance and identified the dead girl's sister as the murderess. When police arrested her, she confessed, but police also arrested Cayce because they felt the only way he could know about the death was by being there. Cayce was finally released.

* Another psychic, Ingeborn Dahl, predicted her own father's death by drowning in Norway and police arrested her for killing him. Fortunately for her, there was not enough evidence and she was released.

The 8-year-old Ghost

There are ghosts and there are ghosts, but Richard Harris feels he has one of the best in an 8-year-old lad who haunts his mansion in England.

Harris, an actor, singer, director and author, has been awakened at 2 A.M. by banging closet doors and the sound of running up the tower stairs. Finally, in frustration, he built a nursery with toys at the top of the stairs and now he feels the ghost is better behaved.

He says he knows it is an 8-year-old because old records reveal a boy was buried in the tower.

STONE MAN OF THE **CATHEDRAL**
VERDEN (ALLER) GERMANY
A TREASURER OF THE CATHEDRAL EMBEZZLED THE CHURCH FUNDS AND WHEN APPREHENDED DENIED IT FROM THE CHURCH WINDOW AND CALLED UPON THE DEVIL TO TAKE HIM IF HE WERE LYING. HE SUDDENLY VANISHED—AND IN HIS PLACE APPEARED A STONE IMAGE.

SPOOKS IN COURT: THE FAMOUS £1000 COLLEY-MASKELYNE LIBEL CASE.

A PRINCE OF SCIENCE FOR THE PLAINTIFF:
DR. ALFRED RUSSEL WALLACE,

Who declares that he had witnessed similar
phenomena to those which Archdeacon Colley
claims to have witnessed.

MR. MASKELYNE'S EXPERIMENT—STAGE 1: MISS CASSIE BRUCE (MRS. E. A.
MASKELYNE) BEGINNING TO EMERGE FROM HER FATHER-IN-LAW'S SIDE.

In virtue of his having performed this illusion mechanically, Mr. Maskelyne claimed £1000 from
Archdeacon Colley.

AN AUTHOR FOR THE DEFENCE:
MR. DAVID CHRISTIE MURRAY.

Who told the Court that Mr. Maskelyne's experiment fairly represented that described in Archdeacon Colley's book on spiritual manifestations.

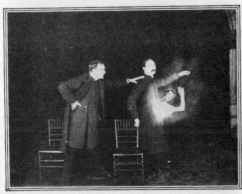

MR. MASKELYNE'S EXPERIMENT—SECOND STAGE: A CLEARER MANIFESTATION
OF MISS CASSIE BRUCE.

MR. MASKELYNE'S EXPERIMENT—FINAL STAGE: THE COMPLETE EMERGENCE
OF MISS CASSIE BRUCE.

MR. MASKELYNE'S MECHANICAL EXPOSURE OF MONCK'S "SPIRITUAL" MANIFESTATIONS, WITNESSED BY ARCHDEACON COLLEY.

THE PLAINTIFF, ARCHDEACON COLLEY:
AWARDED £75 DAMAGES AGAINST MR.
MASKELYNE, WHOSE CLAIM FOR £1000
WAS DISALLOWED.

THE PICTURE PRODUCED IN COURT: HOW ARCHDEACON COLLEY SAW "ALICE-CLOTHED-
WITH-A-CLOUD" EMERGING FROM DR. MONCK'S SIDE.

This manifestation was witnessed at a séance in Bloomsbury by Dr. Alfred Russel Wallace, who testified
in court to his belief in it.

MRS. E. A. MASKELYNE RIVALS "ALICE-
CLOTHED-WITH-A-CLOUD": ARCHDEACON
COLLEY'S VISION PERFORMED MECHANI-
CALLY AT ST. GEORGE'S HALL.

An example of controversy surrounding mediums, apparitions and spiritualists as reproduced from
The Illustrated London News, *May 4, 1907.*

The Holy Relic

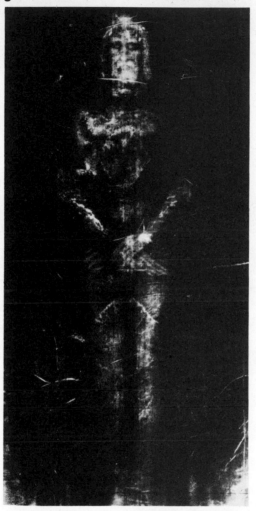

There is one relic in the treasury of the Catholic church which surpasses all others in interest and belief.

The Shroud of Turin is widely believed by Catholics throughout the world to be the burial shroud of Jesus Christ. The popes have never formally pronounced it to be genuine, but Pius XI, speaking as a scholar, displayed a profound belief in its authenticity.

The cloth is a yellowish-stained linen, the color of old ivory, 14 feet 3 inches by 3 feet 7 inches. It carries an image which many say is Jesus, and rivulets on the front and side and across the forehead give the impression of the Crown of Thorns.

The image shows contusions all over the body, and where the left hand meets the wrist, at the base of the palm, there is a vivid mark of blood. The feet appear to be turned towards each other and are partly crossed. A flow of blood emanates from each.

The thorax is expanded and there is a wound in the side. Scientists have studied the cloth and concluded it "may" date back to the time of Christ.

Two photographs of transfiguration medium Mrs. Bullock in 1935, as she changed into the likeness of a Chinese person.

Spiritualist meeting in a Paris drawing room: communicating with "the other side" by means of a hat, table turning, and the pendulum.

The Medium is the Message

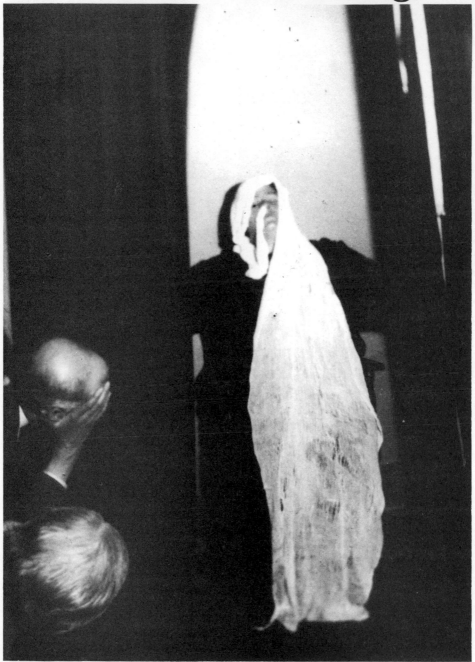

Infra-red photograph of so-called ectoplasm issuing from Helen Duncan, taken by medium Harry Price in 1931 and shown to the jury in his 1944 trial when he was sentenced to 9 month's imprisonment under the British Witchcraft Act.

Medium Famous

A levitation seance in Hamburg.

Not all amazing feats performed by mediums involve crime. The following deeds defy chance and belief.

* Daniel Douglas Home of Scotland once had six people sitting around a table in broad daylight in 1853 when he called up an extra arm and hand. One of the sitters shook hands with it and then turned around while he summoned up several violins and an accordion which played "Home Sweet Home." He could also call up earthquakes and shake a whole house.

* Uri Geller, the Israeli mentalist, snapped a steel fork and bent a key on BBC television in 1974. Minutes after the show had ended, viewers called up from around the country reporting their keys had been bent, their flatware had been crushed and—in one case—a watch that hadn't worked in 25 years had started ticking again!

* The BBC tried to trap clairvoyant Gerard Croisent in 1974 by bringing him a letter which requested a seat at church. Croisent, a Dutchman, predicted everything in the note —even though he didn't read English and was handed a copy sealed in plastic.

THE **PROPHET** OF **DOOM**

Ned Pearson GRAVEDIGGER of Grimsby, England, FOR A PERIOD OF **22 YEARS** ALWAYS VISITED THE HOME OF HIS NEXT CLIENT **24 HOURS BEFORE THE PERSON DIED!**

Life after Death?

(Left to right) The Beatles: John Lennon, Ringo Starr, George Harrison, Paul McCartney; Gerry and the Pacemakers: Gerry Marsden, Freddy Marsden, Les Chadwick, Les McGuire; manager Brian Epstein; Billy J. Kramer and the Dakotas: Robin McDonald, Mike Maxfield, Billy J. Kramer, Ray Jones, Tony Marsfield.

The Beatles—Paul McCartney, George Harrison, Ringo Starr and the late John Lennon—always believed in some form of life after death and in the powers of a medium.

When their manager Brian Epstein died, the Beatles felt he was sending them messages from beyond the grave. They then met with a medium and say that he brought them Epstein's voice and a brief message.

To this day, they are convinced they spoke with their manager and have never publicly revealed the medium who brought him back.

Unknown Power

For years, there have been tales of deathbed visions or visits from dead people. But a systematic investigation of these stories did not come until Dr. Karlis Osis, a noted parapsychologist, personally studied 35,000 deaths.

The findings were stunning. A total of 1,318 deathbed cases reported seeing visions, and 52 percent of the apparitions represented dead persons who were known to the patients.

More than 240 respondents reported views of heaven in wondrous beauty and brilliant color. Another 753 reported that tranquillity replaced pain.

By the Numbers

A college student used mathematics to prove that Christ was the Messiah.

Heather Nadelhoffer presented a paper to the Associated Colleges of Chicago and took seven messianic prophecies from the Bible: that the Messiah would be living between 29 and 33 A.D., be descended from David, be born in Bethlehem, have hands and feet pierced, be tried and condemned, be betrayed by a friend, be sold for 30 pieces of silver.

Using probability methods, she calculated the random chance of fulfilling seven prophecies out of seven was .0000000000000-000000000000000000000000006615.

The Vision

The life of Teresa Neumann baffled even the Catholic church.

Born in 1898, she became bedridden after an accident to her back in 1918 but claimed she was cured by praying to St. Teresa of Lisieux in 1925. The next year, she saw visions of Christ's crucifixion and went into a trance, bleeding from her hands, feet and side.

She also bled from her forehead as if she wore the Crown of Thorns. So deep were her trances and flows of blood that hundreds of people flocked to see her in Konnersreuth, Germany.

Eat and Sleep

Whatever our beliefs, we all need to eat, drink and sleep. Right? Wrong. Some people with the strongest beliefs eschewed both food and rest, challenging science to prove they defied the laws of chance and credibility:
* Margaret Weiss went without food for three years, Katherine Binder of Palatinate went nine years. Liduine of Schiedam lived from 1395 to 1414 on beer and apple peelings.
* British history shows a woman in Westmorland in the 17th century who went for 14 years without eating anything except a spoonful of milk each day before noon.
* Mollie Fancer of Brooklyn, who died in 1916, used to go 10 or 12 weeks without eating and proved to doctors she could read a book without ever opening it.
* Napoleon liked to get by on four hours of sleep a night and Edison on even less. But they were topped by Paul Kern of Budapest who announced in 1920 he had not slept in 22 years!

The Teacup Syndrome

Time now for a break and a good strong cup of tea. But wait, can that cup show the future, our secrets and our shameless past? Chance or not, many believe it can.

We turn now to the tea readers and we leave it to you to determine whether it is science, luck, prophecy or chance. First, we hold our empty cup in the left hand, swirl it slowly in a counterclockwise motion three times and then turn the leaves over into a saucer.

Behold, it reveals:
* The tree shape means good luck, fertility, happiness and a long life.
* A star means the same, only more so.
* A butterfly means success.
* An apple means long life.
* A heart is love, of course.
* Clouds are problems.
* A dog is a faithful friend.
* A cross is suffering and bad luck.
* An arrow is bad news.
* An egg is fertility.

Modern Miracles?

Lourdes, a small French town in the foothills of the Pyrenees, is the most magnetic place of pilgrimage on the face of the earth.

Each year, more than three million people visit the shrine compared with only 1.5 million who visit Mecca. More than 5,000 people claim they have been cured of illnesses and the medical community in Lourdes says 64 cases so defy science that the only term for them is "miracle."

It all started in 1858 when a teenager named Bernadette Soubirous met a woman who called herself Aquero and told her she was the mother of Christ. She told the young girl to go to the spring and wash in it, to tell the priests and people to come in procession and to tell the church to have a chapel built there.

Since then, the accounts of miraculous cures have grown and grown. Miracle or chance? Judge for yourself:

⁎ Michael Gallagher, a 43-year-old administrator with the Internal Revenue Service in Washington, was told he had incurable cancer in 1979 and had only six months to live. He went to Lourdes, and in June 1980, tests showed no sign of the cancer.

⁎ Vittorio Micheli, a 22-year-old soldier with the Italian Alpine Corps, was admitted to hospital with a malignant tumor near his hip. The cancer destroyed most of his iliac bone and caused dislocation. In May 1963 he went to Lourdes and was immersed in the baths encased in plaster from his pelvis to his feet. A week later, his pain subsided and his destroyed bones grew back.

Laying on of Hands

Chance or luck? Belief or coincidence? We cannot judge, but we offer these examples of people who said they had the gift and who brought about the cure:

⁎ Kenneth Hebblethwaite of Bedford, England, was visited by a man who was discharged from the Royal Air Force because of a serious neck disorder. Hebblethwaite put his hands on the man's neck and improvement was instantaneous. The man, Dennis Betts, returned to the air force for a checkup and they pronounced him completely fit.

Priest and devotees at the Miraculous Grotto, Lourdes.

⁎ Harry Edwards deals only with elderly patients suffering from rheumatism and back pains. So effective were his cures, however, that he had to hire 60 typists to cope with his requests for help in Britain in 1971.

⁎ Finbarr Nolan, the seventh son of a seventh son in Ireland, made a fortune faith healing there but moved to England in 1974. He appeared on television to prove the prophecy that worms shrivel in the hand of a seventh son of a seventh son. His career ended when the worms wiggled away happily...

⁎ Srikanta Rao, a fire-walking yogi, announced he would recreate Christ walking on water. More than 5,000 gathered in Bombay in 1966, some paying up to $100 for ringside seats, to see him do it. But Rao stepped into the Ganges and sank like a stone. No refunds, however.

⁎ Police investigated the Family Church of Jesus, which operated in Britain in 1974, after two people died attempting to walk on the North Sea.

Skip to the Beat

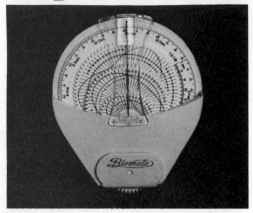

Chance or not, thousands of people believe in biorhythms—the chart from birth to death that measures the internal cycles of physical, emotional and intellectual well-being. Using a series of charted curves, an expert can predict good and bad days if he has a person's birth date, time, latitude and longitude.

But how accurate are these predictions? In the case of Clark Gable, they proved fatal.

A Swiss importer named George Thomman was being interviewed on radio in New York on November 11, 1960, about the science of biorhythms. He worked out Gable's chart because the actor was in hospital recovering from a heart attack while filming *The Misfits*.

Thomman looked ahead and predicted Gable would have to be very careful on November 16 because his physical rhythm would be critical. Gable suffered a second heart attack on that day and died.

Charts of the Famous

We turn now to other people and their charts and wonder if their lives were ruled by chance or biorhythms:

* Marilyn Monroe died of an overdose of pills on August 5, 1962, a day on which both her physical and emotional charts were at their lowest.
* Judy Garland died of a drug overdose June 21, 1969, when her chart showed a critical emotional low, plus physical and intellectual lows.

* Swimmer Mark Spitz won his record-breaking seven gold medals at the 1972 Olympics on days his physical chart showed highly favorable readings.
* TWA flight 514 crashed into a mountain outside Dulles Airport in Washington, D.C., on December 1, 1974—a day that showed critical physical readings for the air controller handling the flight, the pilot and the copilot.
* On November 3, 1973, a Pan Am jet crashed at Logan Airport in Boston, killing three crew members including the pilot and copilot. Both had critical physical rhythms.
* President Gerald Ford granted former president Richard Nixon a full pardon on September 8, 1974, a day when Ford's chart showed negative intellect and emotions.
* Senator Ted Kennedy was involved in the infamous car accident at Chappaquiddick on July 18, 1968, a day when he was emotionally and physically critical.
* President Anwar Sadat of Egypt declared war on Israel on October 6, 1973, a day when he was intellectually high and emotionally low.
* Physical lows were also recorded for assassins: Lee Harvey Oswald on November 22, 1963, when he killed President John F. Kennedy; Jack Ruby on November 23, 1963, when he killed Oswald; and Sirhan Sirhan on June 5, 1968, when he killed Robert F. Kennedy.

The Sporting Life

Evel Knievel on his famous Harley Davison motorcycle.

There's no argument that a major sports achievement requires outstanding physical prowess, but biorhythm charts show athletes can also be suffering from emotional and intellectual highs or lows on big days:

* Benny "The Kid" Paret was knocked out by Emile Griffith on March 24, 1962, a day when all his systems were critical. He died 10 days later, the next all-critical day on his biorhythm.

* Studies show that in 1927 Babe Ruth hit 13 of his record-breaking 60 home runs on days when he was at physical or emotional peaks.

* Evel Knievel nearly died in his attempt to rocket across the Snake River Canyon on September 8, 1974, when his physical rhythm was critical.

* Franco Harris of the Pittsburgh Steelers made the "catch of the century" to put his football team into the playoffs on December 23, 1972, when all three rhythms were at a high.

THE **PRESENTIMENT THAT PROVED PROPHETIC**
HÉLÈNE BOUCHER (1908-1934)
THE FAMED FRENCH AVIATRIX HAD A SUPERSTITIOUS FEAR OF THE FIGURE "30" AND REPEATEDLY REFUSED TO COMPETE IN ANY RACE ON THE 30th DAY OF A MONTH
PERSUADED TO FLY ON NOV. 30, 1934, SHE WAS KILLED IN A CRASH FROM A HEIGHT OF 300 FEET

Accident-prone

Are accidents ruled by chance, the unlucky break? Maybe not, if we judge by these studies of biorhythm charts.

Prof. Hans Schwing of Zurich looked at the birth date of 700 people who had suffered serious accidents and found that 401 of the accidents had occurred on singly, doubly or triply critical days.

And Col. Wolfgang Kornback of the Swiss flight-training center analyzed 130 student pilot crashes and found that 70 percent happened when the pilot's rhythms were critical.

A study by the Japanese police force showed that 59 percent of 1,166 traffic accidents happened on critical days. The Workmen's Compensation Board in British Columbia, Canada, analyzed accidents and found that 13,284 occurred at critical times.

Black Cats and All That

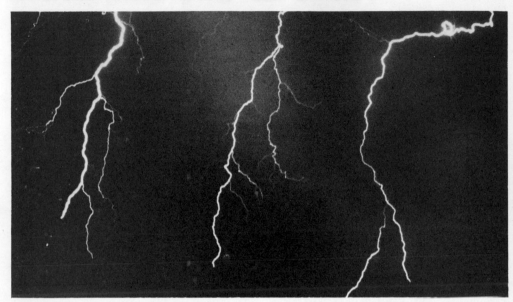

We all know the familiar ones, but each country or civilization throughout the world seems to have its own particular type of superstition. Luck, chance, fate were all rooted in curious fears or beliefs dating back thousands of years.

Lightning, for example, has always been personified as a deity or seen as a manifestation of the wrath of God. Ancient Babylonians showed God as a figure with a boomerang for thunder in one hand and a spear for lightning in the other.

The Persians believed lightning was an indication of divine anger, striking down cattle, men and whole forests. In ancient Greece, a spot struck by lightning was sacred forever. The Greeks also believed thunder on the right side the night before a battle was good luck.

With all these fears, it was no wonder various ancient civilizations developed talismans against lightning. In Scandinavia, burning the Yule log at Christmas is said to protect the house and occupants for a year.

Inhabitants of Shropshire, England, used to believe a piece of hawthorn cut on Holy Thursday protected both a house and the people inside. Other European countries believed lightning would not harm you if you hid the scissors, covered all mirrors, kept away from wet dogs and horses, and laid down on a feather bed.

American Indians believed thunder was caused by the wings of great birds, while some Negroes in the American South were convinced that if lightning flashed while a man was dying, the Devil had come for his soul.

Best of all was an old Maryland tradition that if you got a splinter off a tree struck by lightning and shaped it into a toothpick, you could cure a toothache almost instantly.

8 **GREYHOUNDS** CROSSED THE FINISH LINE *IN THE SAME ORDER AS THEIR POST POSITIONS!*
7th Race, Jacksonville (Fla.) Kennel Club – Jan. 23, 1950

Superstitious Sports

Athletes are almost as bad as actors when it comes to superstitions. Did you know, for example, that it is considered unlucky for a baseball player to spot a cross-eyed woman in the stands?

On the other hand, noticing a red-haired woman usually means a big day—especially if the player can get her to give him a hairpin. Gloves are also considered lucky—if they're left pointing to the home-team dugout when a player goes to bat.

A dog walking across a field is unlucky and all players agree that no season should ever begin on a Friday.

Some other superstitions:

＊ Boxers believe it unlucky if they see a hat lying on a couch or bed before a fight, and any fighter who doesn't spit on his gloves is daring chance.

＊ Cricket players trust to a lucky bat and believe the ball should not be rubbed on certain parts of their clothing. Any bowler who has to restart his run is cursed and any batsman who takes guard twice at the same end will soon be bowled out.

＊ Soccer players believe strongly in a good-luck mascot and many clubs use children. A dressing room tradition calls for the oldest person on the team to bounce a ball to the youngest. If he catches it on the first bounce, luck will hold. And most players still lace the left boot first, bounce a ball three times before kickoff and keep their wives or girl friends away from the game.

＊ Golfers like to carry an old club that they no longer use in their bag. They also believe it is unlucky to approach a tee from the front, change a club after selection or clean a ball when they're ahead. In teeing off, it's a good omen to use a ball with a 3 or a 5 on it, and golfers hate to start at 1 P.M., the 13th hour of the day.

＊ Jockeys hate to see their boots standing on the floor before a big race—they should always be kept on a shelf.

＊ Yachtsmen believe it unlucky to take first place in any practice race for a major competition. And if they are becalmed, wind can always be raised by scratching the mainmast.

Show Business

Ancient superstitions are nothing compared with the ones that have grown in the theater. Even today, actors and actresses religiously follow certain beliefs, unwilling to trust a play or movie to chance:

＊ It is bad luck to quote Shakespeare backstage when anyone else's play is being performed.

＊ Actors never use anything new on opening nights—especially shoes, which might cause a fall.

＊ If your shoes squeak on the first entrance, legend has it the audience will love you. And if you stumble on a line, you will gain audience sympathy.

＊ But it is bad luck to whistle backstage or even in a dressing room. And if you spill powder during makeup, you've got to dance on it.

＊ It is the kiss of death for a play if you say the last line during dress rehearsal. And the curtain must never fall after the last act of a dress run-through.

＊ It's unlucky to open an umbrella on stage in the United States, but not in Britain. But both countries agree yellow, black and green are unlucky because yellow was the color of medieval devils, black is too funereal and green has a corpselike color under lights.

＊ Real coffins must never be used, but all money and jewels must be genuine. Also, never use a real cross on stage or make the sign of the cross.

＊ Ballet dancers always spit on a shoe ribbon before dancing, consider lilacs unlucky and won't let anyone touch a wig once it's in place.

Ignoring Fate

How effective are those theater superstitions? Judge for yourself.

Actress Bea Lillie broke with superstition and sang the same verse of a song twice while with an American touring company in Chichester, England.

Without warning, a huge arc light crashed to center stage.

Hanging by a Thread

Strips of colored rag hung from fence and bushes around Cloutie Well, Scotland, for good luck.

Flip a Coin

It's a practice that dates back to the ancients, but was really defined in the time of Julius Caesar.

His head was on every Roman coin, and so powerful was his rule that decisions were made according to whether a tossed coin came up heads—a practice that continues today with the old saying "heads you win, tails you lose."

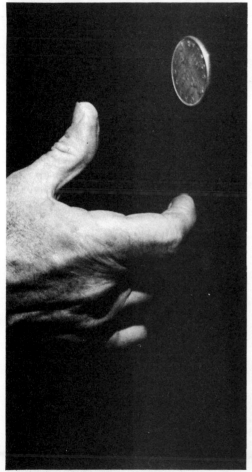

He Had the Edge

A professor of statistics was describing the law of chance to his students and illustrated his lecture by tossing a coin. It came to rest vertically on its edge and the odds against this happening were calculated at 1,000,000,000 to 1.

Around the House

Athletes and actors don't have a monopoly on superstitions—there are some dandies around the house. For example, you're daring chance if you are visiting a house and put a chair against a wall when you leave. Legend has it you will never come again.

And:

* It is a bad omen to place a bed across the floorboards—it should be parallel to the boards.

* You'll sleep better if the head of the bed is pointed north, but it's unlucky to put the left foot on the floor first when you awake.

* Stumbling on the way upstairs can be a sign of a wedding or other good luck.

* Pins should always be picked up because they'll bring you good luck. And if you blow out a candle accidentally, there will be a marriage.

* If three people make a bed, there will be a death within a year. And you should always get out of bed on the side you got in to avoid getting up on the wrong side.

* It's very unlucky to cut hair on Good Friday, and passing a person on stairs without touching or speaking is tempting fate.

Lighter than Air

A 1936 photograph of a yogi poised in mid-air with no visible support except for a draped stick. Witnesses to the event swore as to its authenticity.

Call it chance, fate, luck, superstition or kismet, but history is riddled with stories of uncanny predictions. Monarchs had personal astrologers, businessmen hired fortune-tellers, and wars were won and lost because of prophecies.

We turn now to true stories of forecasting, eerie tales where a man or woman looked into the future with incredible accuracy.

THE DALAI LAMA
ruler of Tibet
A PRISONER OF THE RED CHINESE IN HIS OWN PALACE, ANNOUNCED HE WOULD ESCAPE ON THE AFTERNOON OF MARCH 17, 1959 — THAT AFTERNOON, ALTHOUGH RED TROOPS SURROUNDED THE PALACE AND HUGE SEARCHLIGHTS WERE TRAINED ON THE BUILDING, *THE DALAI LAMA AND 80 COMPANIONS ESCAPED — UNDER COVER OF A SUDDEN SANDSTORM*

Looking at War

Sir Stafford Cripps, the British ambassador to Russia, forecast the date of the German attack on Russia with uncanny accuracy.

On April 23, 1941, he predicted that the Germans would attack on the morning of June 22, which they did. German documents captured after the war show that Hitler did not set the date until April 30.

Albert Purdue, on May 4, 1945, in Chicago, predicted the war in Europe would end on May 8 at 8 P.M. He was right.

> ### WILLIAM LILLY
> NOTED ASTROLOGER — 1602–1681
>
> **WAS HIRED BY OLIVER CROMWELL TO TRAVEL WITH THE ARMY AND ENCOURAGE THE SOLDIERS WITH PREDICTIONS OF VICTORY!**
> *LILLY PREDICTED THE GREAT LONDON FIRE OF 1666*

The Houses of Death

Is this the Borley Rectory ghost? The photographer saw nothing, but these shadows appeared in the negative.

We all know of tales of haunted houses and cursed ships, but even stranger are the tales of buildings that carry prophecies of doom:

* A gypsy forecast that anyone who barred the doors of the parish church in Odstock, England, would die. Two men defied the prophecy and both died violently, so the church warders hurled the keys of the door into a nearby river.

* Rosenstein Castle near Stuttgart, Germany, was completed in 1829 by King William I of Wurtemberg, but he did not use it for 35 years because a fortune-teller predicted he would die there. In 1864, at age 83, he finally moved into the castle and died five days later.

* East Riddlesden Hall in Bingley, England, carried on its facade a motto that the Murgatroyd family would live there as long as the River Aire flowed beside it. In 1692 the Aire suddenly changed its course and the Murgatroyds were evicted.

* A sorcerer sentenced to death predicted that the day he died, the spire would bend at the church of St. Sulpice in Fourgeres, France. He was burned at the stake and the spire tilted that same day almost 18 inches off center.

* King Christian IV of Denmark dedicated the tower of St. Katherine's Church and was told by the builder that he would endure as long as the building. The tower was demolished in a storm on February 28, 1648, and King Christian died that same night.

* Lady Anne Grimston, on her deathbed in 1731, said seven trees would grow from her grave. Soon after she was entombed, the seven sprouted.

* The two gate posts of the stately home of Margam in Wales were left standing for centuries because of a prophecy that if they were destroyed, the Mansell family would vanish with them. In 1744, Lord Thomas Mansell pulled down the pillars and within a year he died with all his brothers.

Lucky Dream

Seaman Thomas King refused to sail on the bark *Isidore* from Maine because he dreamed it would be wrecked—with seven corpses on its deck. The *Isidore* foundered 10 miles from port, and when the wreck was boarded, seven corpses were found.

Tales of Doom

Even stranger are the stories of people who tried to escape fate, to cheat a prophecy of death, or who tempted fate by making their own predictions:

* King Ferdinard V of Spain never visited Madrigal, his wife's birthplace, because a soothsayer told him he would die in Madrigal. He fell ill in a little village in 1516, learned it was called Little Madrigal and died of fright.

* King Edward IV of England was warned by an astrologer that a brother with the initial G would murder Edward's two sons. Edward drowned his brother George, but his sons were killed by his older brother—the duke of Gloucester.

* King Henry IV of England was warned by a seer that he would die in Jerusalem. In 1413 he died of fright after an attack of epilepsy when he learned he had been carried into Westminster Abbey's Jerusalem Room.

* Aeschylus, the father of Greek drama, was warned by a prophet he would die of a blow from heaven. The poet was killed when an

eagle mistook his head for a rock and dropped a turtle on him in 456 B.C.

* George Thaler predicted in 1643 he would die five years later at 4 A.M. on September. 4. He died of natural causes exactly the way he had forecast it, and his prophecy is engraved on his tombstone.

* Mrs. Albert E. Royce told her family on December 3, 1874, that she dreamed she would become sick on November 27, 1877, and die on the following December 3. She sickened and died in Bowling Green, Ohio, exactly as she had envisioned it.

* Anne Cutler of Charlestown, Massachusetts, on her deathbed in 1683, said: "My son, Robert, will go with me whither I am bound." Nearly 2,000 miles away in Barbados, her son Robert died at exactly the same time.

Ship Story

A boy saved his brother's life by a dream.

Adrian Christian dreamed five times that he was captain of a ship and his brother Thomas was on another ship that was sinking So real was the dream that Adrian's mother recorded it in the family Bible.

Forty-seven years later, in 1880, Adrian was captain of a ship and had the same dream again, this time with the word "family" on a piece of paper. He altered the course of his ship, taking it into dangerous waters, and sighted another ship in distress.

Adrian rescued 269 people including his brother Thomas. And the sinking ship was named the *Family*.

An Emperor Believes

Diocletian was told by a Druid in 270 that he would become the emperor of Rome by killing a wild boar.

At first, Diocletian believed him and killed many boars. But the prophecy did not come true.

Then in 284 he stabbed to death the assassin of Emperor Numerianus and succeeded him. He learned later the man he killed was named Aper, the Latin word for wild boar.

The Prophetic Curse

Sir Anthony Browne, the owner of Cowdray House, evicted tenants from properties given him by King Henry VIII. He heard them prophesy that his home and family would be destroyed by fire and water.

In 1793 it came true. The house was burned to the ground and the last male heir drowned in the Rhone a week later. Three cousins succeeded to Sir Anthony's estate— but all drowned while still childless.

A Cardinal Fear

A prophecy frightened a man to death.

Cardinal Thomas Wolsey was warned in 1515 that Kingston would mark the end of his life. He avoided the town in Britain for many years, but died of shock when King Henry VIII sent a man to arrest him in 1530.

The constable's name was Kingston.

The Book of Success

Prophecy is also recorded in politics and business.

The 1892 yearbook of a small high school in Lynn, Massachusetts, carried the prediction that student Frank G. Allen would become governor of the state.

Exactly 37 years later, it happened.

Another yearbook in Newton, Illinois, predicted in 1913 that Frank E. Martin would become vice-president of the Illinois Central Railroad and return to Newton for a reunion. Again, 37 years later, it happened to the letter—or number!

Bad Dreams

Rev. Charles Morgan of Winnipeg, Canada, dreamed of an old hymn he had not heard in years. In the background, he heard the sound of voices and rushing water.

He felt compelled to ask his congregation to sing it, including the line "we pray to thee for those in peril on the sea."

The date was April 14, 1912, and the *Titanic* was sinking.

The Death of Socrates.

All the News

A drunken reporter predicted a disaster before it happened.

While sleeping off a binge, Byron Somes, a reporter for the *Boston Globe*, had a dream about a volcanic eruption on a tropical island. So impressed was he that he wrote an account of the disaster.

He left the dream story in his desk. Only later was it learned the story was a full account of the explosion of Krakatoa in 1883, and every detail in Somes's story was later confirmed.

The Hymn That Wrote Itself

In 1861, during the Civil War in the United States, Julia Ward Howe was visiting Washington and saw soldiers returning from the war.

She went to sleep for the night, but when she woke at dawn, she found herself sitting at her desk. She discovered she had written a poem of five verses entitled "The Battle Hymn of the Republic."

To her death, Julia wondered if she really wrote the masterpiece or was just an "instrument" of something bigger.

Other Dreams

Dreams have offered solutions for seemingly insoluble problems and have helped make history. People have written their best works in dreams and many authors kept pen and pencil at their bedsides:

* Samuel Taylor Coleridge dreamed the great poem "Kubla Khan" during a drug-induced sleep. He said it was originally between 200 and 300 lines but he stopped recording it after 54 because the dream "melted away."
* Socrates was told in a dream that he must write his work on paper. He passed this on to his student, Plato, because the dream came the night before he was ordered to drink hemlock.
* Alexander the Great attacked and conquered Tyre because he dreamed of a satyr. His oracle told him *sa* in Greek meant "yours" and *tyr* meant the city of Tyre. Before the dream, Alexander was planning to withdraw his army.
* William Blake decided to use copper-plate etching for his manuscripts after his dead brother, Robert, appeared in a dream and told him it was the answer to printing his works.
* Giuseppe Tartini, an 18th-century composer, dreamed he sold his soul to the Devil and then handed him his violin. The Devil played a sonata and when Tartini awoke, he wrote his most famous piece—the *Devil's* Sonata.
* Loewi, a German physiologist, won the Nobel Prize for his discovery that nervous impulses in the body were created by chemical transmissions. He admitted later that the experiment came to him in a dream.
* Elias Howe dreamed of a group of hunters who told him he would be dead if he didn't invent a machine that could sew. Howe did it and his sewing machine needle was designed after the spears the hunters carried.
* Two days before his assassination, Lincoln dreamed of a funeral in the White House.

Vintage Poe

Edgar Allan Poe wrote *The Narrative of A. Gordon Pym* in 1838, which told of three shipwrecked survivors in an open boat who killed and ate a cabin boy named Richard Parker.

Nearly 50 years later, three shipwrecked survivors really did kill and eat a cabin boy. And his name was Richard Parker.

The Book's Curse

Morgan Robertson wrote a novel entitled *The Wreck of the Titan* in 1898, which corresponded in striking detail to the sinking of the *Titanic* in 1912.

The *Titan* had its maiden voyage in April, the same month the *Titanic* went down. The *Titan* was also a big fast luxury ship out to beat a record, had many wealthy people aboard, and was in the North Atlantic when it hit an iceberg.

Other coincidences: the *Titan* carried 3,000 passengers (*Titanic* 2,207); it had only 24 lifeboats (*Titanic* 20); it was traveling at 25 knots (23 knots); it weighed 75,000 tons (66,000); it was 800 feet long (882.5); and it had three propellers, the same as the *Titanic*.

The End of the World

History is full of stories about people who predicted the end of the world, but they haven't been right yet!

* The first rumor that the end of the world was imminent spread over Europe just before 1000 A.D. So many religious people believed it was about to happen that the construction of places of worship was abandoned. New cathedrals and churches sprang up only after the fateful period had passed, and many charitable foundations in Europe date from those terrible days of fear.

* On June 13, 1857, Paris was full of terrified people who fled to St. Denis to escape being buried alive under collapsing houses because all knew the world was ending.

* William Miller made 3,200 speeches predicting the end of the world and then announced to his American followers that he would speak no more because the Day of Judgment was coming on March 14, 1844.

* Les Spangler, a Pennsylvanian prophet, fixed October 1908 as the month the world would end.

* Michael Stiefel, a Lutheran in Germany, spent his leisure time making mysterious mathematical calculations. One day he roundly announced that the end of the world would come October 3, 1533, at exactly 10 A.M.

* Johan Stofler, a professor of mathematics in Germany, studied the stars and announced that Saturn, Jupiter and Mars would be dangerously close to Earth in 1524 and cause a second flood. Thousands believed him: debtors refused to pay back loans, people hoarded food, warehouses were looted, rich people moved to the mountains and others built barges and kept them near their houses.

FIEND, FROM DÜRER'S DESCENT OF CHRIST INTO LIMBO.

The Great Disappearing Act

Thousands of articles, books and studies have been made of the area of the Atlantic Ocean bounded by Florida, Puerto Rico and Bermuda. Some say most of what has happened there can be explained, others say the disappearances of ships and aircraft far outweigh the laws of chance.

Does some beast lurk beneath the depths as some have suggested? Is the area ruled by strange magnetic forces? Is this the one area of the world where man was never meant to travel?

We turn first to the story of the ship *Mary Celeste*, which set sail from New York to Genoa in 1872.

It was found one month later in the Triangle with all sails set. But the crew had disappeared and the ship was totally deserted. The only lifeboat was missing but there was a full supply of food and fresh water and the ship wasn't sinking.

Money and clothes had been left behind, the forehatch was open, there was an ax mark on the ship's rail and blood was found on the deck. Violence, mutiny, piracy or the curse of the Triangle?

Nothing changes the fact, however, that between 1800 and 1972, more than 40 cases of mysterious disappearance were registered. Among other tales:

* In October 1492 Christopher Columbus and the crew of the *Santa Maria* reported a great flame of fire in the middle of the Triangle and severe compass disturbances.
* Charles Lindbergh was flying over the Triangle in the *Spirit of St. Louis* in 1928 when both his compasses malfunctioned.
* A month later, a tri-engine Fokker was misdirected when the compass varied 50 percent. The plane subsequently crashed.

* In November 1943 a B-24 was flying over the area in good weather when it suddenly went out of control and the crew bailed out.
* A veteran seaman aboard the *Atlantic City* saw the automatic steering device suddenly turn his ship in a full circle. Simultaneously, a ball of lightning appeared.
* Don Henry, captain of the salvage tug *Good News*, was beset with engine failure in 1966. When he went on deck, he looked behind him in clear weather and noticed the barge he was towing had disappeared, even though the line was still taut. It slowly rematerialized.
* Passengers aboard the *Queen Elizabeth 2* had to transfer to another ship in April 1974, when the huge liner suddenly reported a mysterious power failure in the middle of the Triangle.
* The crew of the U.S. Coast Guard cutter *Hollyhock* saw a large land mass appear on radar in the middle of an empty sea. Checks of the radar showed no faults.
* Photographer Dr. Jim Thorne took pictures of an electric storm in the Triangle in July 1975. When he developed the film, the picture also showed an old square-rigged sailing ship although nothing had been sighted. The apparition still has not been explained.
* A Beechcraft Bonanza plane flew into a huge cloud, lost radio contact and after four minutes suddenly found itself over Miami— with 25 gallons more fuel than it had over the Triangle.
* A National Airlines 727 was lost on radar for 10 minutes. When it landed, it was found that all watches on board and the ship's chronometer had lost exactly 10 minutes.
* An Eastern Airlines flight suffered a major bump over the Triangle and landed at a nonscheduled stop to check damage. All

aboard noticed their watches had stopped at the time of the jolt, and the plane's fuselage was severely burned.

* A Cessna 172 flew directly over the airport tower at Grand Turk Island in the Bahamas, but the tower could not contact the pilot on the radio. They could hear him, however, telling his passenger that "there's nothing below." Pilot and passenger then flew off and were never seen again.

Spontaneous Human Combustion

Medically, it is never discussed, and it has never been observed in animals. But the fact remains that it is one of humanity's biggest mysteries.

Chance, superstition, a curse? We do not know. Scientifically, it is impossible. There is no known way in which burning human tissue can generate the colossal temperatures needed to effect almost total consumption of the body.

Even if such temperatures were possible, they would vaporize almost everything else around. Yet cases exist and beliefs persist:

* Mary Reeser was found in 1951 by a neighbor in St. Petersburg, Florida, burned to death in her chair. She had been reduced to 10 pounds of fine white ash and the chair and a lamp were also destroyed, but the blackened circle was only a yard wide and everything else in the room was untouched.

* The Countess Cornelia di Bandi was found dead in her bedroom in 1763. Her legs and head were untouched, but nothing was between them except a pile of ashes.

* Mary Carpenter was out on a boat with her husband and children near Norfolk, Virginia, in 1938 when she suddenly burst into flames and was reduced to ashes. There was no damage to the boat, the husband or the children.

* Euphemia Johnson, a London widow, was found lying in a heap within her clothing in 1922. All that remained of her was her bones, but the clothing was undamaged.

* George Turner, a truck driver in Britain, was found burned to a cinder inside the cab of his lorry, but a can of gasoline beside him had not ignited.

Balls of Fire

There are thousands of unexplained sightings of fireballs, but scientists still cannot explain the phenomenon. Is it an energy source from within the ball, is it electric, electromagnetic, nuclear or minute fragments from a meteor?

One of the oddest stories on record concerns Prof. R. C. Jennison of the University of Kent in England. An expert in electronics and electricity, Jennison was eminently qualified to make the following observation.

In 1963 he was dozing in his seat aboard an Eastern Airlines flight from New York to Washington. He was the only passenger aboard, although a stewardess was sitting opposite him. Their seatbelts were fastened because the captain had warned of turbulence.

Suddenly, out of the door leading to the pilot's cabin came a glowing ball of fire, 10 inches in diameter and blue-white in color. It hovered above the cabin floor, slowly moved down the aisle and finally disappeared into the rear of the plane. Scientists accept the incident as having actually occurred. Believe it or not!

THE **SKELETON**
OF THOMAS A BECKET
WAS PLACED ON TRIAL BY
KING HENRY VIII ON A CHARGE
OF USURPATION OF OFFICE
365 YEARS AFTER HIS DEATH.
AN ATTORNEY WAS ASSIGNED TO
DEFEND THE ACCUSED, BUT THE
SKELETON WAS FOUND GUILTY

Index

A

Accidents, 153–154, 176, 178–180, 203, 206, 220–221, 224, 229, 297
The African Queen, 24
Age, 182–183
Albatross, 142, 278
Alchemy, 266, 285
Alda, Alan, 14
Alda, Robert, 14
Alexander the Great, 239, 306
Ali, Muhammad, 82
Allegret, Marc, 4
Allyson, June, 5
Alpha Centauri, 124
Alston, Walt, 66
L'Amour, Louis, 34
Ancient technology, 237, 238
Andersen, Lale ("Lili Marleen"), 29
Andrea Doria, the, 89
Andrews, Dana, 23
Anna Karenina, 36
Apocalypse Now, 107
Appearance, 187
Arbuckle, Fatty, 14
Archimedes, 147
Arlington, Lord, 48
Art & artists, 38–41, 103, 106, 172, 187, 209
Aspinall, John (the "king of Clubs"), 42, 55
Assassination, 170, 257, 265, 296
Astaire, Fred, 8
Asteroids, comets & meteors, 124, 127–129, 271
Astrology, 266, 268, 269, 271, 272
Atari, 94
Avalanche, 120, 144
Ayres, Lew, 18

B

Baby Doll, 18
Bacall, Lauren, 4, 278
Baccarat, 54, 56, 96
Bailey, Pearl, 24
Baker, Carroll, 18
Baker, Tom, 24
Ball, Lucille, 20
Bandages, 176, 214
Bankhead, Tallulah, 20
Bankrupts & bankruptcy, 108, 111, 110, 174
Bardot, Brigitte, 4
Barretta, 12
Barnum, P.T., 100
Barrymore, John, 11, 22
Bartlett's *Quotations*, 32
Baseball, 64, 66, 67, 68, 69, 83, 133, 299
Basketball, 66, 68
Battery, 237
Battles, 239–243, 248–250, 253–256
Baum, L. Frank, 21
Beatles, the, 292
Beauty contests, 5
Beethoven, 28
Bell, Alexander Graham, 96, 145
Bergman, Ingrid, 20
Berkeley, Busby, 9
Berle, Milton, 25
Bermuda triangle, the, 266, 308
Bernhardt, Sarah, 42, 56, 100
Best Foot Forward, 5
Bets, 42, 48–49, 56–57
"Big Bopper, the", 30
Bingo, 49
Biorhythms, 266, 296, 297
Birth, 176, 183, 195–200, 229
Birth of a Nation, 12
Blackjack, 42, 44–45, 47, 50, 54
Blake, Robert, 12
Blake, William, 306
Blood and Sand, 14
Bloopers, 22
Blue moon, 144
Bogart, Humphrey, 6, 12, 24
Bolger, Ray, 21
"Bond, James", 9
Bond, Michael, 35

HUMPHREY BOGART
THE FAMED MOVIE TOUGH GUY HAD SUCH SWEET
FEATURES AS A CHILD, THAT HE WAS
USED AS A MODEL TO SELL BABY FOOD

ST. THOMAS à BECKET
(1118 - 1170)
ARCHBISHOP OF CANTERBURY
WAS APPOINTED ARCHBISHOP
ON A *TUESDAY*
WAS PUT ON TRIAL
ON A *TUESDAY*
BANISHED FROM ENGLAND
ON A *TUESDAY*
RETURNED FROM EXILE
ON A *TUESDAY*
SLAIN ON A *TUESDAY* and
CANONIZED ON A TUESDAY

Books, 31–37
   classics, 32
   plays, 32
   poetry, 32
   sci-fi, 36, 37
   westerns, 32
Boone, Pat, 25
Bow, Clara, 5
Bowling, 76
Bowser, 28
Boxing, 64, 68, 82, 83, 119, 299
Bradbury, Ray, 36
Brady, Diamond Jim, 117
Brahms, 28
Brains, 224–225
Brando, Marlon, 22, 272
Brennan, Walter, 7
Bridge, 42, 44, 47
Bridges, Beau, 14
Bridges, Jeff, 14
Bridges, Lloyd, 14
Broadway, 4, 24
Brown, Andy, 72
Brown, Charlie, 272
Browning, Ted, 12
Burglary, 174
Burns, Robbie, 32
Burton, Richard, 4, 12, 272
Burroughs, Edgar Rice, 32
Byron, Lord, 86, 208, 209

C

Caesar, Julius, 238, 301
Caesar, Sid, 63
Caine, Michael, 6
Canadian Mounties vs The Atomic Invaders, 23
Cancer, 176, 178, 205
Cantor, Eddie, 25
Captain Blood, 11, 22
Card games, 42, 44
Carmen Jones, 22
Carnegie, Andrew, 173
Carradine, David, 14
Carradine, John, 14
Carradine, Keith, 14
Carson, Johnny, 7
Caruso, Enrico, 139
Cary, Joyce, 35

JOHANNES BRAHMS
(1833-1897) THE FAMED GERMAN COMPOSER TOOK NEARLY 20 YEARS TO COMPLETE HIS FIRST SYMPHONY — *BUT FINISHED HIS SECOND, THIRD AND FOURTH SYMPHONIES IN QUICK SUCCESSION*

EDWIN BOOTH

THE BROTHER OF JOHN WILKES BOOTH, WHO ASSASSINATED LINCOLN, HAD RISKED HIS OWN LIFE TO SAVE ABE LINCOLN'S SON, ROBERT, FROM BEING CRUSHED BY A TRAIN IN JERSEY CITY, N.J.

Casablanca, 12, 17, 24
Casanova, 276
Cassidy, David, 14
Cassidy, Jack, 14
Cassidy, Shaun, 14
Castle of Kerousein, the, 46
Cat on a Hot Tin Roof, 13
Cellini, Benvenuto, 41
Chagall, Marc, 41
Champion, 278
Chaney, Lon, 18
Chaplin, Charlie, 14, 202
Charms & talismans, 266, 274, 277
Charge of the Light Brigade, the, 256
Charles, I, 247
Charles II, 229, 171, 285
Charlie's Angels, 23
Cheating, 45, 47, 48
Cheevers, Gary, 72
Chess, 42, 56, 57
Chico & the Man, 107
Chopin, 28, 209
Churchill, Winston, 100
Cigarettes, 149
Cimino, Michael, 12
Circus, the, 3, 22
Civil suits, 169
Clark, Bobby, 24
Clift, Montgomery, 22
Cobalt bomb, 215
Cobham, Lady Ann, 48
Cockfighting, 49
Cocktails, 150
Cody, Buffalo Bill, 21, 100
Cohn, Harry "King", 12
Coins, 301
Collectibles, 100, 108
Coleridge, Samuel Taylor, 306
Columbus, Christopher, 244, 308
Como, Perry, 25, 84, 96
Common cold, 210, 211
Computers, 90, 150, 153, 154, 215, 237
Computer crimes, 156, 163, 171
Concepcion, Dave (Cincinnati Reds), 66
Concorde, the, 97
Confessions, 169
Conklin, Chester, 14
Connery, Sean, 9
Connors, Jimmy, 202
Contests, 61, 63
Cooke, Sam, 30

JULIUS CAESAR

THE ROMAN GENERAL AND STATESMAN ASSASSINATED BY A GROUP OF NOBLES RECEIVED 23 STAB WOUNDS -- *BUT* ONLY ONE WAS FATAL

Coolidge, Calvin, 272
Cooper, Gary, 3, 8, 20
Coronation Street, 23
Cosgrove, Jack, 20
Costa, Mario, 5
Costravitskaya, 12
Counters, 47, 50
Craps, 48, 57
Crawford, Joan, 18, 20
Credit cards, 109, 112, 171
Crichton, Michael, 36
Cricket, 299
Crime, 156–175
Crime and Punishment, 31
Croce, Jim, 30
Cromwell, Oliver, 272, 303
Crosby, Stills, Nash and Young, 28
Cuckor, George S., 20
Cures, 204, 206, 210, 211, 213, 220, 229
Curtis, Jamie Leigh, 14
Curtis, Tony, 14, 18
Custer's Last Stand, 256
Cyclones, 135

D

Damn Yankees, 23
Damone, Vic, 25
Dandridge, Dorothy, 22
Dante, 31
Darby O'Gill & the Little People, 9
da Vinci, Leonardo, 39, 41
Davis, Bette, 20, 272
Davis Jr., Sammy, 105
Day, Doris, 12, 26
Dead End, 12
Dean, James, 17
Death, 176, 179, 180, 185, 186, 187, 220
"Death Car, the", 17
Decameron Nights, 22
Deer Hunter, the, 12
Defoe, Daniel, 31
de Havilland, Olivia, 20
de Niro, Robert, 107
Diamonds, 86, 88, 129
Diary of Anne Frank, the, 6
Dice, 42, 48, 57
Dickens, Charles, 36
Diet, 183, 295

THE MAN WHO ORGANIZED THE VOYAGE
THAT DISCOVERED AMERICA
Martin Pinzon (1440?-1493)
A SPANISH NAVIGATOR AND
SAILOR, SECURED THE CREW FOR
THE EXPEDITION OF 1492 AFTER
COLUMBUS HAD FAILED TO DO SO,
HELPED FINANCE IT AND COMMANDED
THE PINTA, THE SHIP THAT FIRST
SIGHTED LAND. *PINZON WAS
THE MAN THE CREWS LOOKED TO
FOR DIRECTION BECAUSE THEY
DISTRUSTED COLUMBUS AS A FOREIGNER*

MARLENE DIETRICH

IN 1933 WAS ASKED TO LEAVE PARIS,
FRANCE, BY THE PREFECT OF POLICE
BECAUSE SHE WAS WEARING TROUSERS

Dietrich, Marlene, 7, 107
Dillinger, John, 272
Di Maggio, Joe, 83
Dinosaurs, 126–127, 234
Dirty Dozen, the, 8
Discoveries and inventions, 90–94, 96–99, 145–150, 176, 212–215, 244, 250, 255
Disease, 176, 182, 183, 205, 206, 208–209, 227
Divine Comedy, the, 31
Dixon, Jeanne, 272
Doctors, 211
"Dr. Who", 24
Dolce Vita, La, 8
Dolphins, 127
Doors, the, 30
Dornhoefer, Gary, 72
Dostoevsky, Feodor, 31
Double Indemnity, 12
Douglas, Kirk, 9, 14, 95, 278
Douglas, Michael, 14, 95
Dracula, 284
Dracula, 12, 18, 34
Dragons, 234
Dreams, 306
Dry cleaning, 146
Duels, 220, 221
Duncan, Isadora, 268
Durocher, Leo, 66
Duryea, Dan, 9

E

Earthquake, 120, 136, 137, 139
Eason, Reeves, 20
Eastwood, Clint, 9
East of Eden, 17
Edsel, the, 97
Ebsen, Buddy, 21
Edison, Thomas, 22, 96, 293
Einstein, Albert, 271
Eisenhower, Dwight, 272
Electromagnetism, 149
Elliot, Mama Cass, 15
Empire Strikes Back, the, 7
Environment, the, 120, 130–144
Epilepsy, 240
Erasmus, 211
Esposito, Phil, 72

DWIGHT D. EISENHOWER WHO BECAME ONE OF THE MOST SUCCESSFUL COMMANDERS IN HISTORY, WAS ADMITTED TO WEST POINT, IN 1911, ONLY BECAUSE THE APPLICANT WHO RANKED AHEAD OF HIM IN HIS ENTRY TEST FLUNKED THE PHYSICAL

ALBERT EINSTEIN (1879-1955) ONE OF THE GREATEST MINDS OF ALL TIME, WHILE ATTENDING ELEMENTARY SCHOOL IN MUNICH, GERMANY, MAINTAINED ONLY AVERAGE GRADES

Ether, 176, 212
Evans, Dame Edith, 4
Exorcist, the, 12
Eyes, 188

F

Fabergé, 108
False teeth, 212
Famous gamblers, 42, 45–49, 56, 59–61
Fairbanks Jr., Douglas, 107
Farewell to Arms, A, 20
Farrow, Mia, 24
Faulkner, William, 37
Feet, 202
Fellini, 8
Feminism, 192
Fields, W.C., 21
Fire, 120, 131, 140, 141, 175, 248
Fireballs, 309
Fisher, Carrie, 14
Fisher, Eddie, 14, 25
Fishing, 80
Fitzgerald, Barry, 9, 14
Fitzgerald, F. Scott, 20
Flemming, Victor, 20
Flynn, Errol, 7, 11, 20, 22, 272
Fonda, Henry, 14
Fonda, Jane, 14
Fonda, Peter, 14
Fontaine, Joan, 20
Football, 64, 66, 68, 80, 83
Ford, Gerald, 202, 272, 296
Ford, Glen, 24
Ford, Harrison, 7
Forgery, 94, 171, 174, 172
Fossils, 234, 235, 238
Fox, Charles James, 46
Francis, Connie, 25
Frankenstein, 18
Franklin, Benjamin, 281
Franklin, Sidney, 20
Fraud, 175
Frazier, Joe, 83
Freaks, 12
Frederick of Germany, King, 32
Freud, Sigmund, 272
Frilled collars, 94
Frisbee, the, 99, 206

*T*HE **FAMED ACTOR WHO OWED MUCH OF HIS SUCCESS TO ADVERSITY W.C. FIELDS** (1879-1946) WHO WAS A VAGRANT AT THE AGE OF 11, GOT HIS SWOLLEN RED-NOSE FROM ALLEY FIGHTS, AND HIS HOARSE VOICE *FROM CHILDHOOD COLDS*

SIGMUND FREUD (1856-1939) THE FOUNDER OF PSYCHOANALYSIS, WHO STRESSED THE SEXUAL FACTOR IN NEUROSES, WAS IN HIS PRIVATE LIFE *EASILY SHOCKED AND PURITANICAL*

From Here to Eternity, 9
Future, the, 151, 153

G

Gable, Clark, 3, 9, 20, 296
Gambling, 42–57
Garland, Judy, 14, 18, 21, 202, 296
Gauguin, Paul, 39, 41
Gehrig, Leo, 3
Geiberger, Al, 70
Gender, 176, 181, 183, 219
Germ cultures, 176, 214
Getty, J. Paul, 278
Ghandi, Mahatma, 272
Ghosts, 286, 287
Giant, 17
Giants, 235
Gilbert, Sir W.S., 29
Gilbert & Sullivan, 29
Gleason, Jackie, 272
Goddard, Paulette, 20
Godfather, The, 117
Godfrey, Arthur, 25
Going My Way, 9
Gold, 86, 87, 88, 89, 285
Goldwyn, Sam, 14, 21
Golf, 64, 70, 74, 83, 299
Gone with the Wind, 3, 20
Goodbye, Columbus, 4
Gould, Elliot, 94
Goya, 106
Grable, Betty, 202
Graduate, The, 12
Grant, Cary, 10
Gratton, Gilles, 72
Great Flood, the, 236
Great Gatsby, The, 12
Great Train Robbery, the, 162
Green, Hubert, 70
Green Pastures, 25
Greenstreet, Sydney, 6
Gretzky, Wayne, 72
Grey, Zane, 32
Griffith, D.W., 12
Guggenheim, Peggy, 103
Guillotine, the, 166
Duke of Guise, the, 46, 245

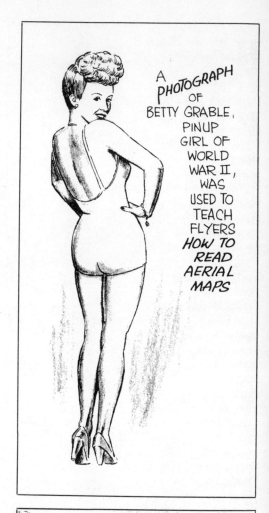

A PHOTOGRAPH OF BETTY GRABLE, PINUP GIRL OF WORLD WAR II, WAS USED TO TEACH FLYERS *HOW TO READ AERIAL MAPS*

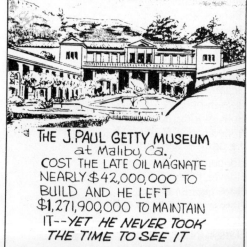

THE J. PAUL GETTY MUSEUM at Malibu, Ca. COST THE LATE OIL MAGNATE NEARLY $42,000,000 TO BUILD AND HE LEFT $1,271,900,000 TO MAINTAIN IT--*YET HE NEVER TOOK THE TIME TO SEE IT*

H

Hail, 131
Hair, 159, 188–189
Haley, Alex, 23
Haley, Jack, 21
Hamilton, Margaret, 21
Handel, 28
Haney, Carol, 5
Hangings, 166
Harlow, Jean, 16, 18
Harris, Richard, 286
Harrison, 292
Harrison, Rex, 202
Hawn, Goldie, 23
Haydn, Richard, 12
Hayward, Susan, 18, 20
Hawks, Howard, 4
Hayworth, Rita, 18
Hearst, Patty, 272, 307
Heaven's Gate, 12
Hefner, Hugh, 102
Height, 188
Heiresses, 105, 106
Hello Dolly!, 24
Hendrix, Jimmy, 30
Hepburn, Katherine, 20, 21
Hercules against the Moon Men, 23
Hickock, Wild Bill, 45
High Noon, 8
High Sierra, 12
Hilton, Conrad, 278
Hippocrates, 269
History, 232–265
Hitler, Adolph, 261, 264, 269, 303
Hoaxes, 172
Hockey, 64, 66, 72
Hogan, Ben, 83
Hole-in-one, 64, 74
Holly, Buddy, 30
Hollywood, 4, 6–7, 9–12, 14, 15, 17, 20–22, 24, 95
Hooper, 7
Hoover, J. Edgar, 272
Hope Diamond, the, 104
Horoscopes, 239, 268, 269, 272
Horror movies, 18
Horses, 249
Hot-air balloons, 148

J. EDGAR HOOVER
chief of the FBI
WAS SCARED BY A BLACK CAT!

*HE STEPPED ON ITS TAIL
IN A RAID ON THE ROGER TOUHY GANG
AND THE CAT'S YOWL WAS
HOOVER'S GREATEST MOMENT
OF FEAR IN **30** YEARS AS HEAD
OF THE G-MEN*
Chicago, Ill. – Dec. 29, 1942

THE GRAVE OF ROBIN HOOD
near Kilkees, England
IS LOCATED ON THE SPOT
*WHERE HIS LAST ARROW
HIT THE EARTH*

Hotdogs, 150
Houdini, Harry, 14
Houseman, AE, 209
Howard, Leslie, 20
Howe, Elias, 306
Howe, Gordie, 72
Howell, Roy (Toronto Blue Jays), 66
How Green Was My Valley, 9
Hudson, Rock, 3, 6
Hughes, Howard, 34
Hula-hoop, the, 99
Humankind, 120, 126, 234
Humperdinck, Englebert, 15, 63
Hurricanes, 120, 134
Hutton, Betty, 5

I

Ice-cream cones, 150
Idiot-geniuses, 191
Immunization, 212
Incubators, 149
Ingenuity, 90–91, 93, 94, 98
Inheritances, 102, 105
Instant coffee, 149
Insulin, 176, 213, 215
Insurance, 106, 107, 167, 228
Intermezzo, 20
Intolerance, 12
Ipcress File, The, 6
Irving, Clifford, 34

J

Jackson, Andrew, 82
Jackson, Kate, 23
Jack the Ripper, 202
Jagger, Mick, 27, 30
Jazz Singer, The, 9
Jefferson, Thomas, 60, 281
Jefferson Airplane, The, 27
Jesus of Nazareth, 23
Jobs, 114, 115, 271
Johnson, Lyndon, 272
Jolson, Al, 9
Jones, Brian, 30
Jones, Jennifer, 18

CZAR IVAN THE TERRIBLE

(1530-1584) of Russia, IN 1584 SAW IN THE SKY A COMET HE CONSIDERED A SIGN OF HIS DEATH, AND THAT SAME YEAR PROPHETS' PREDICTED HE WOULD DIE ON MARCH 18 th
-- *THE DAY HE ACTUALLY SUCCUMBED*

THE PROPHECY OF A PROUD GRANDFATHER LYNDON B. JOHNSON WHO BECAME A U.S. SENATOR AND PRESIDENT WAS BORN ON AUG. 27, 1908, AND ON THAT DAY HIS GRANDFATHER RODE HIS HORSE AROUND STONEWALL, TEXAS, SHOUTING: *"A UNITED STATES SENATOR HAS BEEN BORN TODAY"*

Jones, Shirley, 14
Jones, Tommy Lee, 4
Joplin, Janis, 29
Jourdan, Louis, 22
Joyce, James, 37
Jung, C.G., 277
Jungle Book, The, 33
Junk mail, 111
Juveniles & crime, 164

K

Karloff, Boris, 18
Keaton, Buster, 14
Keats, John, 208
Kelly, Emmett, 22
Kelly, Grace, 13
Kennedy, Jacqueline, 105, 272
Kennedy, John F., 232, 265, 266, 272, 296
Kennedy, Robert, 272, 296
Kennedy, Ted, 272, 296
"Keystone Kops, The", 11, 12
King, Carole, 28
Kipling, Rudyard, 33
Khrushchev, Nikita, 272
Knievel, Evel, 266, 297
Koufax, Sandy, 202
Kojak, 8
Kozinski, Jerzy, 32
Kramer vs Kramer, 23
Kubla Khan, 89, 306

L

Ladd, Jr., Alan, 14
Lady Chatterley's Lover, 34
Laemmle, Carl, 12
Landis, Carole, 5
Landin, Michael, 18
Laugh-In, 23
Laughter, 176, 206
Law, the, 156, 168, 175, 180
Lawrence, D.H., 34
LeBoeuf, Armand, 28
Left-handedness, 202
Leigh, Janet, 14
Leigh, Vivien, 20

JOHNNY WEISSMULLER
THE SCREEN'S MOST FAMOUS TARZAN,
WAS HIRED FOR THE ROLE
WITHOUT A SCREEN TEST

THE ACTOR WHO BECAME A STAR
WITHOUT UTTERING A LINE
Boris Karloff
ACHIEVED STARDOM IN HIS
ROLES IN "FRANKENSTEIN"
AND "THE OLD DARK HOUSE"
--*YET HE NEVER SPOKE
A WORD IN EITHER*

Lennon, John, 30, 292
Leob, Gerald, 278
Lerner, Alan Jay, 26
Levis, 92
Lewis, Jerry, 26
Life Savers, 90
Lightning, 132, 140, 245, 298
"Lili Marleen", 29
Lincoln, Abraham, 232, 265, 266, 272, 281, 306
Lindbergh, Charles, 308
Little Eva, 28
Little Foxes, The, 9, 107
"Little Tramp, The", 14
Loewi, 306
Lollabrigida, Gina, 3, 5, 18
Lombard, Carole, 5, 20
Longest Yard, The, 7
Longfellow, H.W., 281
Loren, Sophia, 5, 8
Lost Horizon, 12
Lotteries, 42, 49, 58, 59, 60, 61, 277
Lourdes, 295
Love Story, 4, 117
LSD, 146
Luck, 278
Lucky days, 266, 274
Lugosi, Bella, 18, 23
Lynnyrd, Skynnyrd Band, the, 30

M

Mabel's Strange Predicament, 14
MacGraw, Ali, 4
Mack, Ted, 25
MacLaine, Shirley, 5
MacMurray, Fred, 12
Magicians, 266, 183, 185
Mail-order, 90
Maltese Falcon, The, 6, 12, 24
Mamas and the Papas, The, 15
Mansfield, Jayne, 15
Manson, Charles, 32
Mao Tse-tung, 272
Marriage & divorce, 192, 194
Martin, Dean, 26
Marvin, Lee, 9

HENRY W. LONGFELLOW
1807 - 1882

WROTE **18** VOLUMES
OF POETRY,
WAS GRADUATED FROM
BOWDOIN COLLEGE
AT **18**, MARRIED HIS
SECOND WIFE **18** YEARS
LATER, WAS A PROFESSOR
AT HARVARD
FOR **18** YEARS,
*AND WAS STRICKEN WITH
HIS FATAL ILLNESS ON
MARCH 18, 1882*

Marx, Groucho, 3, 20
M*A*S*H, 94
Mastroianni, Marcello, 8
Matisse, Henri, 41
Mayer, Louis, B., 12
Mazarin, Cardinal, 46
Mazarin, the Duchess of, 46
McCartney, Paul, 202, 292
McDonald's, 93, 104
McEnroe, John, 202
Mead, Margaret, 272
Medical miracles, 222–223, 226
Medical Research, 216–218
Mediums, 291–292
Mencken, H.L., 272
Menzies, William Cameron, 20
Merrick, David, 24
Merrily We Go to Hell, 23
Merv Griffin Show, The, 28
Messigny, Louis Maria, 26
Michelangelo, 41
Microfilm, 149
Microscope, 214, 215
Midwives, 196
Millionaires, 84, 90, 91, 92, 98, 99, 102, 104, 105, 106
Mills, Hayley, 25
Milton, John, 31
Minnelli, Liza, 14
Miracles, 266, 295
Misfits, The, 296
Mistaken identity, 163–164
Mitchell, Margaret *(Gone with the Wind)*, 20
Mitchum, Robert, 21
Mona Lisa, The, 39
Monday, 186
Monet, Claude, 38, 41
Money, 84–119
Monkees, The, 28
Monopoly, 91
Monroe, Marilyn, 5, 12, 18, 268, 279, 296
Moon, Keith (the Who), 105
Morgan, Frank, 21
Morrison, Jim, 30
Morse, Samuel, 146
Movies, 4–24, 26, 36, 94, 107, 117
Movie stars, 3–24, 25, 94, 95, 107, 296
Mozart, 28
Multiple births, 198
Murder, 156, 158, 159, 160, 162, 164, 165
Musial, Stan, 83

GROUCHO MARX

THE GREAT COMEDIAN MADE HIS FIRST PROFESSIONAL APPEARANCE AS A BOY SOPRANO IN A CHURCH – –

BUT WAS DISMISSED FOR PUNCTURING THE ORGAN WITH A HATPIN

WOLFGANG MOZART (1756–1791)

ONE OF THE WORLD'S FOREMOST COMPOSERS, DIED WHILE CREATING A REQUIEM MASS FOR THE DEAD

Music, 25–30
    classical, 26–29
    opera, 29
    rock, 27–30
My Fair Lady, 26
My Favorite Martian, 23

N

Namath, Joe, 80
Napoleon, 253, 254, 255, 272, 279, 293
Nature, 120, 130–144
Newman, Paul, 9
Newton, Sir Isaac, 147
Newton-John, Olivia, 25
New Yorker, The, 35
Nicholson, Jack, 9, 95
Nicklaus, Jack, 74
Niven, David, 18
Nixon, Richard, 272, 296
Noise, 180
Northmore, MP, William, 46
Norton, Ken, 82
Nostradamus, 276
Novak, Kim, 5, 202
Noxcema, 98
Nuclear energy, 149
Nuclear war, 185
Numerology, 279–281, 297
Nureyev, Rudolf, 12

O

Octopus, 88
Odds, 21, 24, 45, 37, 44–45, 50, 57–60, 63,
66, 68, 74, 76, 78, 81–82, 111, 112–115,
119, 150, 158–159, 164, 178, 185–186,
198, 200, 203, 301
Oil, 87
Olaf of Norway, King, 48
Olof of Sweden, King, 48
Olympics, the, 68
One Flew Over the Cuckoo's Nest, 95
One Million B.C., 5
On the Waterfront, 22
Organized crime, 161
Orr, Bobby, 72

MARSHAL NEY COULDN'T SAY "NO"!
MARSHAL MICHEL NEY (1769-1815)
ORDERED TO BRING NAPOLEON BACK IN CHAINS AFTER
BONAPARTE'S ESCAPE FROM ELBA, WAS EXECUTED
FOR TREASON BECAUSE INSTEAD OF ARRESTING
NAPOLEON HE SUCCUMBED TO HIS ARGUMENTS
AND JOINED IN HIS MARCH ON PARIS

AN **OCTOPUS** MOVES
FORWARD BY WALKING
AND ALWAYS SWIMS
BACKWARDS

O'Sullivan, Maureen, 24
Oswald, Lee Harvey, 296

P

Pablum, 215
"Paddington Bear", 35
Paganini, Niccolo, 208
Page, Patti, 25
Paige, Satchel, 69
Paint Your Wagon, 9
Palmistry, 266, 282
Panama Hattie, 5
Paper, 146
Paradise Lost, 31
Pasteur, Louis, 176, 212
Patents and trademarks, 90, 94, 96, 119
Patton, 17
Pearl Harbor, 262
Pearson, Lester B., 173
Peck, Gregory, 24
Peg O'My Heart, 5
Pele, 64, 79
Perfect hands, 44, 47
Petrified Forest, The, 24
Pet rocks, 94
Pets, 176, 211, 228
Photocopiers, 149
Photography, 149
Picasso, Pablo, 40
Pickford, Mary, 18
Pidgeon, Walter, 8
Pilgrims, 247
Pink lemonade, 3
Pinochle, 44
Piquet, 46
Pitcairn Island, 201
Placebo, 210
Places to live, 118
Planets, 266
   Earth, 120, 122, 236
   Jupiter, 124, 236
   Mars, 124, 129, 236
   Neptune, 269
   Pluto, 269
   Saturn, 124, 236
   Sun, the, 122

PABLO PICASSO
ONE OF WHOSE PAINTINGS RECENTLY
SOLD FOR $3,000,000, WAS SO POOR
EARLY IN HIS CAREER THAT HE BURNED
SOME OF HIS DRAWINGS TO KEEP WARM

MARCO POLO (1254-1324)
THE VENETIAN EXPLORER OF CHINA,
TAKEN PRISONER WHEN HE WENT
TO WAR AGAINST GENOA IN A
SHIP FITTED OUT AT HIS OWN
EXPENSE, WHILE A CAPTIVE
WROTE A BOOK ABOUT HIS
JOURNEYS THAT WAS SO WIDELY
ACCLAIMED, HIS CAPTORS
GAVE HIM HIS FREEDOM (1298)

Uranus, 269
Venus, 124, 236
Plato, 238, 306
Playboy, 102
Player, Gary, 70, 74
Pleyel, Ignace, 26
Plummer, Christopher, 6
Poe, Edgar Allan, 32, 307
Poker, 42, 44, 45, 47
Polaroid, 96
Pollock, Jackson, 103, 106
Polo, Marco, 34
Pool, 82
Pools, football & soccer, 49, 58, 59, 78
Porpoise, 142
Potato chips, 150
Potatoes, 155
Pound, Ezra, 32
Preminger, Otto, 94
Presley, Elvis, 25
Press, the, 4, 8, 9, 23, 27, 306
Pressure suits, 215
Pride of the Yankees, The, 9
Prince Charles, 23, 271
Prince Rainier of Monaco, 13
Prinz, Freddy, 107
Prophecies, 238–241, 266, 292, 304-305, 306
Psychics, 266, 286
Predictions, 27, 57, 136, 137, 150–151, 263, 271, 276, 291, 303, 306–307

Q

Queen Elizabeth I, 59, 118, 269, 271
Queen Victoria, 100, 268
Q-tips, 98

R

Races,
dog, 49, 298
horse, 42, 49–53, 57, 68, 133, 299
Radio vacuum tubes, 94
Raft, George, 12
Raging Bull, 107
Raiders of the Lost Ark, 7

THE CONCERT THAT SAVED 175,000 LIVES!

SHAH QULI, SUMMONED TO ENTERTAIN SULTAN MURAD IV OF TURKEY, WHO HAD ORDERED THE EXECUTION OF EVERY INHABITANT OF CONQUERED BAGHDAD, PLAYED HIS PSALTERY SO MOVINGLY THAT *THE BRUTAL SULTAN BROKE INTO TEARS - AND OFFERED THE MUSICIAN ANY REWARD WITHIN HIS POWER.*

SHAH QULI ASKED FOR AND WAS GRANTED - *THE LIVES OF HIS 175,000 FELLOW CITIZENS OF BAGHDAD*

BABE RUTH
ARRESTED FOR SPEEDING, WAS SENTENCED TO A DAY IN JAIL, BUT WAS RELEASED IN TIME TO PLAY THE LAST 3 INNINGS OF A GAME, *AND GIVEN A POLICE ESCORT TO THE BALL PARK*

Rain, 130, 133
Rameses II, 269
Rasputin, 183
Reagan, Ronald, 18, 24
Rebel Without a Cause, 17
Records, 57
Redding, Otis, 30
Rembrandt, 106
Renoir, 41
Reynolds, Burt, 7
Reynolds, Debbie, 5, 14
Riders of the Purple Sage, 32
Ringling Brothers' Circus (Barnum & Bailey), 22
Robberies, 164, 165, 173, 175
    unsuccessful, 167
Robin Hood, 22
Robin Hood, 242
Robinson, Edward G., 24
Robinson Crusoe, 31
Robots, 152–153
Rockefeller, David, 102
Rodin, Auguste, 38
Rodriguez, Chi Chi, 70
Rogers, Will, 34
Rolling Stones, The, 27, 30
Rooney, Mickey, 18
Roosevelt, Franklin, 272
Roots, 23
Roulette, 50, 54
Rubies, 86
Ruby, Jack, 296
Ruth, Babe, 83, 202, 297

S

Sabrina, 107
Saccharin, 147
Sadat, Anwar, 296
Safety pins, 94
Sagan, Francoise, 37
St. John of the Cross, 209
St. Thomas Aquinas, 269
St. Valentine's Day Massacre, 100
Sand, George, 28
Sanders, Colonel, 24, 115
Sandwich, the Earl of, 47
Sasquatch, the ("Big Foot"), 235
Satan, 245

A **HARVARD STUDENT**
IN A DEBATE ON THE
ANNEXATION OF HAWAII IN 1902,
STRESSED THAT PEARL HARBOR
*COULD PLAY AN IMPORTANT
ROLE IN THE U.S. NAVY !*
39 YEARS LATER PEARL
HARBOR'S NAVY ROLE WAS
FORCIBLY CALLED TO THE
ATTENTION OF THAT STUDENT

-- FRANKLIN D. ROOSEVELT

Savalas, Telly, 8
Schiller, Friedrich, 208
Scipio Africanus, 22
Scott, George C., 14
Seberg, Jean, 9
Sellers, Peter, 107
Selznick, David O., 12, 20, 279
Selznick, Lewis J., 12
Selznick, Myron, 20
Sennett, Mack, 11, 12, 14
Sex, 194
Shakespeare, William, 31
Sha-na-na, 28
Shaw, George, 83
Shaw, George Bernard, 272
Sheik, The, 14
Sheridan, Anne, 20
Shipwrecks, 142, 143, 278
Show business, 2–30, 299
Shrapnel, 94
Shroud of Turin, the, 288
Sydney, Sylvia, 23
Silver, 86, 89, 108
Simenon, Georges, 36
Sinatra, Frank, 9, 14, 272
Sinatra, Nancy, 14
Sirhan, Sirhan, 296
Sleep, 186, 293
Sleuth, 6
Smokey & the Bandit, 7
Smuggling, 170
Snapshots, 149
Snead, Sam, 74
Snow, 131
Sobers, Gary, 202
Soccer, 63, 78–79, 133, 299
Socrates, 306
Solar heat, 90
Sondergaard, Gale, 21
Sothern, Ann, 18
Soul, David, 28
Sousa, John Philip, 26
South Pacific, 23
Space food, 154
Spanish Armada, the, 246, 269
Spark plug, 237
Spencer, Lady Diana, 23, 271
Spic and Span, 98
Spitz, Mark, 296
Spontaneous human combustion, 266, 309
Sports, 64–83, 297, 298

ROBERT
LOUIS
STEVENSON
(1850-1894)

FOUND THE
PLOT FOR
"DR. JEKYLL AND MR. HYDE
IN A
NIGHTMARE

Spy in Your Eye, 23
Stakes, 46, 48, 49, 57, 60, 245
Stamp, Terence, 202
Stamps, 172
Standard time, 94
Starfish, 215
"Star of Africa, The", 86
Starsky and Hutch, 28
Star Wars, 7
Starr, Ringo, 15, 292
Stealing, 156, 158, 162, 164, 166, 170, 173, 175
Steamboats, 94
Steiger, Rod, 17
Sterilization, 176, 213
Stevens, Cat, 28
Stevenson, Adlai, 272
Stevenson, Robert Louis, 208
Stewart, James, 9
Stevens, Andrew, 14
Stevens, Stella, 14
Stills, Steven, 28
Stock market, 116, 117, 173
Stoker, 34
Strasberg, Lee, 12
Streep, Meryl, 23
Streetcar Named Desire, A, 8
Streisand, Barbra, 6
Stress, 229
Sugar, 118
Suicide, 187, 250
Superman, 100, 107
Superstitions, 50, 64, 66, 69–70, 72, 74, 80, 82, 266, 273, 284, 298–299, 301
Sutherland, Donald, 94
Swindlers, 86, 172
Synthetic blood, 216

T

Taft, W. Howard, 272
Tartini, Giussepe, 306
Tarzan, 21, 32
Tate, Sharon, 32
Tattoos, 159
Taylor, Elizabeth, 13, 18, 107, 272
Taxes, 115, 250
Tea, 93
Tea-leaves, 293

THE **TOY POODLE** OWNED BY GEORGE SAND, THE FRENCH NOVELIST, WAS THE INSPIRATION FOR CHOPIN'S FAMED MINUTE WALTZ, *THE TRUE NAME OF WHICH IS* **THE WALTZ OF A LITTLE DOG**

MARK TWAIN

MADE A FORTUNE AS A WRITER, BUT LOST IT ALL AS AN UNSUCCESS-FUL **INVENTOR**

Telephone, 96, 145
Telegraph, 146
Television, 7–8, 12, 18, 23–25, 28, 107, 149
Tennis, 68
Theater, 4, 6, 14, 299
Thoreau, Henry David, 35, 208
Thunder Road, 21
Thunderstorms, 133, 144
Thurber, James, 35, 100, 272
Tiant, Louis, 66
Tidal wave, 137, 142
Time-twins, 266, 271
Titanic, the, 84, 88, 103, 142, 306, 307
Tolstoi, Leo, 36, 209
Tonight Show, The, 7
Tork, Peter, 28
Tornados, 135
"Toto", 21
Towering Inferno, The, 9
Treasure,
    buried, 86, 89, 94
    sunken, 84, 88, 89
Triskaidekaphobia, 281
Truman, Harry, S., 202, 259, 272
Tuberculosis, 208
Tuboeuf, 46
Turner, Lana, 6
Turpin, Ben, 22, 107
Twain, Mark, 32, 272
20,000 Leagues Under the Sea, 9
Twins, 199–200
$2.00 bill, the, 119
Two Little Bears, 23
Typewriter, the, 97
Typhoons, 135

U

UFOs, 124, 125
Unitas, Johnny, 83
Universe, the, 120, 122, 125
Uranium, 87
Utrillo, Maurice, 41

V

Valens, Richie, 30
Valentino, Rudolph, 14

TOM THUMB

THE 2-FOOT-11-INCH MIDGET
EXHIBITED BY P.T. BARNUM IN THE
1860s, WAS CONSIDERED SO
CUTE THAT HE WAS *KISSED
BY 2,000,000 WOMEN—
INCLUDING THE QUEENS OF
ENGLAND, SPAIN, FRANCE
AND BELGIUM*

VIRGIL
(70-19 B.C.)

THE ROMAN POET ONCE PAID THE
EQUIVALENT OF $100,000 TO PROVIDE A
LAVISH FUNERAL FOR A HOUSEFLY
*--AND BURIED IT IN A
SPECIAL MAUSOLEUM ON HIS ESTATE*

Vallee, Rudy, 22
Valley of the Dolls, 18
Van Eyck, 38
Van Gogh, Vincent, 38, 209
Vengeance, 167
Verne, Jules, 37, 268
Virgil, 238
Visconti, Luchino, 8
Visitors from outer space, 236
Volcano, 136, 137, 139
Voltaire, 59, 164
Vukovich, Pete, 66
Vulcanized rubber, 147

W

Wagner, Richard, 268
Wallenda, Karl, 22
Walpole, Horace, 48
Walston, Ray, 23
War, 232, 242, 243, 245, 246, 251–252,
255–256
Wars,
the Hundred Years War, 242
the American Revolution, 250
the French Revolution, 251–252
War of 1812, 254–255
Civil War, 255–256
Crimean War, 256
World War I, 257–260
World War II, 260–264
Warhol, Andy, 106
Warner, Jack, 56
Warner, Sam, 9
Washington, George, 60, 212, 250
Water, 122, 140, 141
Watson, Tom, 70
Wayne, John, 3, 7, 18, 22
Weather, 186
Weber, Karl Maria von, 208
Webster, Daniel, 255
Webster, Noah, 31
Weight, 189–191
Weight Watchers, 93
Weissmuller, Johnny, 21
Welch, Racquel, 8
Wellman, William, 20
Wells, Charles ("The Man Who Broke the
Bank"), 55

THE **ADMIRALTY BUILDING** in London WAS BUILT FROM THE PLANS *FOR A LUNATIC ASYLUM* IN 1887. ITS ARCHITECT WAS SUDDENLY SUMMONED TO SUBMIT THE ADMIRALTY PLANS TO QUEEN VICTORIA, BY ERROR, SHOWED HER HIS DESIGNS FOR AN ASYLUM — AND AFTER HER ENTHUSIASTIC PRAISE WAS AFRAID TO REVEAL HIS MISTAKE

WALT WHITMAN (1819-1892) AUTHOR OF THE AMERICAN *CLASSIC* "LEAVES OF *GRASS*" ENDED HIS FORMAL *SCHOOLING* AT THE AGE OF **11**

Whales, 127, 142
What's My Line?, 24
Whist, 44
Whitman, Walt, 32
Wild One, The, 22
Williams, Edy, 107
Williams, Emlyn, 4
Wills, 102, 104
Wilson, Dooley, 17
Witches, 245, 266, 275, 284
Wizard of Oz, The, 21
Wood, Sam, 20
World population, 184
World's wealthiest countries, 106
Worst movies, 23
Writers, 31–37, 208, 209
Wrong Box, The, 22
Wyeth, Andrew, 272

X

X-rays, 146

Y

Young, Brigham, 272
Yorkshire Ripper, the, 163

Z

Zanuck, Daryl F., 56
Zanuck, Richard, 94
Zebra in the Kitchen, 23
Zeffirelli, Franco, 23
Zippers, 149
Zombies on Broadway, 23
Zukor, Adolph, 12

ABORIGINAL "X-RAYS"

WITCH DOCTORS AMONG THE AUSTRALIAN ABORIGINES BELIEVED THEY COULD PRODUCE RAIN BY PAINTING ON TREE BARK "X-RAY" ILLUSTRATIONS OF A KANGAROO--*CENTURIES BEFORE DR. ROENTGEN DISCOVERED X-RAYS*

DARRYL F. ZANUCK
(1902-1979)
THE MOVIE MAGNATE BEGAN HIS CAREER AS AN EXTRA FOR ONE DOLLAR A DAY COSTUMED AS AN *INDIAN MAIDEN*

Photograph Credits

Ann Ronan Picture Library: pp.59,96,100
121,123,127,128,129,130,133,136,138,
145,147,148,151,168,197,207,209,212,
213,214,224,238,244,270,276,282,286,
289,309.

Associated Press: p.89.

B.B.C. Hulton Picture Library: pp.36,56,86,
88,96,119,126,139,141,146,147,150,209,
225,233,234,239,240,242,243,248,251,
252,254,255,256,257,259,264,265,267,
274,276,283,284,293,301,304,305,306,
307,308.

Bellerophon Books: p.31

Canadian Press: pp.83,87,97,235.

Embassy Pictures Corp.: p.8.

Historic New Orleans Collection: pp.58,99.

Holly Fisher Photographer: p.93.

Illustrated London News Picture Library:
pp.117,184,224,257,261,264,287,293,302.

International Portrait Gallery: p.31

Keystone Press: p.160.

King Features Syndicate: pp.10,23,67,185,
226,227,263,264.

London Features International Ltd.: p.27.

Metropolitan Toronto Police Museum:
pp.158,159,162,164,165,167,168,170,172,
174,175,179,218.

Miller Services: pp.32,102,258,272.

Mirror Newspapers: p.52.

NASA: p.113.

National Film Archive: pp.7,95,296.

National Portrait Gallery: pp.33,34,37,46,
48,208,250,253.

Octopus Books: pp.25,79,122,124,256,262,
269.

Paul Cannon Photographer: pp.91,155,178,
179,203,206,225,301.

Popperfoto: pp.24,38,39,40,41,47,50,54,55,
59,100,101,103,107,108,109,110,125,134,
149,152,157,161,162,163,182,183,191,
192,201,202,208,215,217,222,241,245,
246,247,249,253,260,277,275,278,283,
285,288,289,290,291,292,294,295,297,303.

Public Archives of Canada: p.85.

Robert Estall Photographer: pp.236,300.

R.K. Pilsburg Photographer: p.298.

Royal Canadian Air Force: p.128.

Spectrum Colour Library: pp.3,77,87,97,
112,114,153,188,191,194,237,273.

Steve Back Photographer: pp.43,92,118,
180,202.

St. Michael's Hospital, Toronto: pp.216,228.

Sunday Telegraph: p.104.

Toronto Blue Jays Baseball Team: p.65.

Toronto Star Syndicate: pp.6,70,71,72,73,75,
76,78,79,80,82.

United Press International: p.4.

Universal Pictorial Press: p.187.

Warner Bros.: p.15.

Zoological Society of London: p.142.

With special thanks to Paul Cannon
Photographer for photographs taken
exclusively for this book.